Seeking Symmetry

Seeking Symmetry
Finding Patterns in Human Health

Niall T McLaren Galloway MB FRCS(Ed)
Medical Director, Emory Continence Center
Associate Professor of Urology, Emory University,
School of Medicine, Atlanta, Georgia, USA
Former Chairman, National Association for Continence
Former President, Georgia Urological Association

with Sarah McArthur Smith PhD
Director: The Well-Suited Word, LLC
Atlanta, Georgia, USA

Forewords
Stephen Levin MD FACS
John Sharkey MSc

HANDSPRING PUBLISHING

EDINBURGH

HANDSPRING PUBLISHING LIMITED
The Old Manse, Fountainhall,
Pencaitland, East Lothian
EH34 5EY, Scotland
Tel: +44 1875 341 859
Website: www.handspringpublishing.com

First published 2018 in the United Kingdom by Handspring Publishing

ISBN 978-1-912085-11-8

British Library Cataloguing in Publication Data
A catalogue record for this book is available from the British Library

Library of Congress Cataloguing in Publication Data
A catalog record for this book is available from the Library of Congress

Notice
Neither the Publisher nor the Authors assume any responsibility for any loss or injury and/or damage to persons or property arising out of or relating to any use of the material contained in this book. It is the responsibility of the treating practitioner, relying on independent expertise and knowledge of the patient, to determine the best treatment and method of application for the patient.

Commissioning Editor Sarena Wolfaard
Project Manager Morven Dean
Copy-editor Susan Stuart
Designer Kirsteen Wright
Illustrators Kirsteen Wright; Paul Rodecker, Savannah College of Art and Design; Angus Galloway, Georgia State University
Indexer Aptara, India
Typesetter DSMSoft, India
Printer Replika, India

The Publisher's policy is to use paper manufactured from sustainable forests

Contents

Dedication

To my wife Sian, and to all my many teachers,
thank you for challenging me.

Foreword by Stephen Levin

Humans have been fascinated with symmetry for the entirety of our recorded history, with a fascination traceable all the way back to prehistoric cave paintings and, later, the Greeks. In every religion, every period of art, every architectural movement, you can find symmetry enmeshed in its interstices, often as a driving force that underlies a discipline's foundational principles. My bookshelf holds four books with the word 'symmetry' in the title and several more that have symmetry as an underlying theme, permeating nearly every page. So why a new book about symmetry?

As both a pelvic surgeon and the co-founder and director of the Emory Continence Center, Niall Galloway has developed a unique perspective on the multileveled health problems that require his talents as a surgeon. Other urologists refer patients to his multidisciplinary center for the kinds of complex incontinence problems, pelvic organ prolapse, and reconstructive surgery that have proved challenging for them to resolve or repair, not only from around Atlanta but around the nation.

But how does this work qualify him to write a new book on symmetry? Besides the obvious reality that his surgeries take him into the bilateral inner world of, for instance, two kidneys, a multidisciplinary medical approach requires innovative thinking. And Niall's creative perspective has brought him to biotensegrity, a way of understanding the body not as a conveniently packaged assemblage of component parts but as an evolutionary process in which symmetry plays a major role.

The building block of biotensegrity, what I will call the 'structural generator', is the icosahedron, the most symmetrical of the Platonic solids, and the mechanism of icosahedrons is based on tensegrity (Figure 1). It has been recognized as the structure of most viruses since the 1960s,[1] and it is the heart of DNA structure[2] and cell-to-organism construction.[3-6] A regular icosahedron has 60 rotational (or orientation-preserving) symmetries, and a symmetry order of 120, including transformations that combine a reflection and a rotation. There are ten 3-fold axes, each of which passes through the centers of two opposite faces, six 5-fold axes, each of which passes through two opposite vertices, and fifteen 2-fold axes, each of which passes through the midpoints of two opposite edges (Figure 2).[7]

The twelve vertices of the icosahedron, so central to biosentegrity, can be understood as three mutually perpendicular golden rectangles (Figure 3), a geometric shape whose significance and relevance to the human body Niall has spotted (as he explains in the first chapter). Icosahedrons seem complicated in other hands, but Niall makes sense of their relevance to human health, as he places symmetry at the heart of biological constructs.

What makes this book on symmetry so important is that symmetries are not treated as merely pleasing design patterns, mathematical abstractions important only for their aesthetic qualities. Instead, symmetry appears in its rightful place as a fundamental quality of life, at every scale and at every organizational level. Symmetries lie at the heart of growth and form, evolution, hormonal function, body mechanics, neurological interactions, and the physiology of body functions, and Niall extends our understanding of symmetry to our interactions with agricultural management and nutrition and beyond. He teaches us to look for symmetries in our lives, as essential and central to our well-being.

Just when we think that everything that can be written on a subject has already been written, along comes a new twist that proves us wrong. In *Seeking Symmetry*, Niall Galloway has given us such a book, bringing new insights to this timeless subject.

Stephen M Levin MD FACS
Ezekiel Biomechanics Group
Washington, USA

References

1. Zandi R, Reguera D, Bruinsma RF, et al (2004). Origin of Icosahedral Symmetry in Viruses. Proceedings of the National Academy of Sciences of the United States of America 101: 44.

2. Dage L, Wang M, Deng Z et al (2004). Tensegrity: Construction of Rigid DNA Triangles with Flexible Four-arm DNA Junctions. Journal of the American Chemical Society 126: 8.

3. Levin, SM (1981). The Icosahedron as a Biologic Support System. 34th Annual Conference on Engineering in Medicine and Biology, Houston TX.

4. Levin, SM (1982). Continuous Tension, Discontinuous Compression: A Model for Biomechanical Support of the Body. Bulletin of Structural Integration 8(1): 31-33.

5. Levin, SM (1989). The Icosa-dynamical Model for Biomechanics - the Space Truss. In: Ledington PWJ (1989). 33rd Annual Meeting of the International Society for the Systems Sciences, Edinburgh.

6. Levin, SM (1986). The Icosahedron is the Three-dimensional Finite Element in Biomechanical Support. In: Dillon JR (1986) Society of General Systems Research, 30th Annual Meeting.

7. Hargittai, I (1992). Fivefold Symmetry. Singapore: World Scientific.

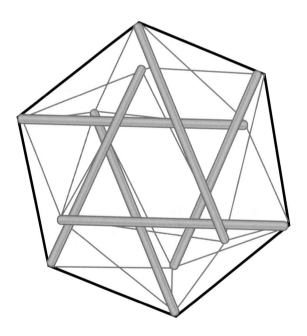

Figure 1

Tensegrity icosahedron.

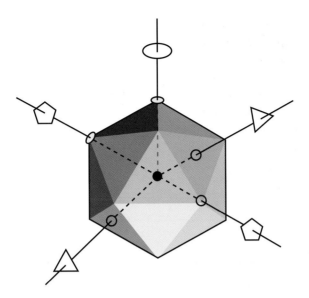

Figure 2

2-fold ⬭, 3-fold △, 5-fold ⬠,
axis of rotation of icosahedrons.

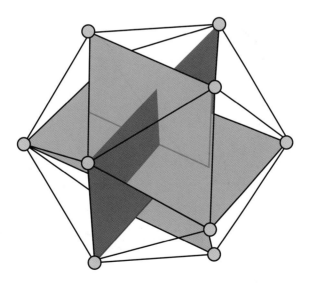

Figure 3

Golden Rectangles in an icosahedron.

Foreword by John Sharkey

'For me, it is far better to grasp the Universe as it really is than to persist in delusion, however satisfying and reassuring.'

- Carl Sagan, The Demon-Haunted World

An innovative thinker and man of the Italian Renaissance, Leonardo da Vinci lacked the insights he would have derived from modern scientific technologies such as electron microscopy or MRI technology. Da Vinci had only the poor light of candle to help illuminate his fertile mind, his unrelenting curiosity and his insatiable hunger for knowledge. Borrowing from, and adding to, ancient wisdom from minds now lost to antiquity, Niall Galloway skillfully guides our journey, relating often-unforeseen points of connection, synergies and symmetries either purposely dismissed or mistaken, through lack of appreciation or understanding, much to our detriment.

This is a timely and critically important work penned by a visionary doctor, surgeon, anatomist, husband, family man, fellow human being, and champion of biotensegrity. Dr. Galloway demystifies nature's predilection for the basic building plans and materials required for animate and inanimate forms. Once armed with this information the reader can translate this knowledge, from the formation of falling snowflakes to the developing embryo, from the pupils of our eyes to the invisible forces within our cosmos (inner and outer). Galloway eloquently describes the symmetry of our nature and indeed the symmetry of all nature.

Dr. Galloway invites us to consider the model of tensegrity as an essential aspect of human architecture (biotensegrity). He combines tensegrity principles with observations concerning the golden section, Fibonacci numbers, and pairs of complementary opposites. These are key points for the reader, as the nature of mathematics and the mathematics of nature drives the architecture of all living and non-living structures. I regularly refer to biotensegrity as 'the model that binds' or 'anatomy for the 21st century'. This book accurately reflects this centrality of biotensegrity as 'symmetry-seeking in practice, applying it to understanding the human body, to demystifying some aspects of medicine, and to looking at industrial systems that most directly affect human health: big food, big pharma, and agribusiness'.

Concerned agencies, organisations, and individuals can act upon the undeniable realities described by Galloway's vision to provide scientific, research-based evidence for the adverse effects of widespread inappropriate use of antibiotics. His perspective provides a new alternative in the effort to avoid further catastrophic health problems in the future.

Biotensegrity does not exclusively concern the internal forces acting to ensure the integrity of cells, groups of cells, tissues, organs, organ systems, and organisms. It involves everything that influences the external and internal environments, which in turn dictate biotensegral integrity. A book dedicated to expressing this global view is long overdue. We are fortunate to have in Niall Galloway someone who shares da Vinci's cross-disciplinary fascination with forms, which together with his experience as a modern physician and researcher makes him uniquely qualified to offer this viewpoint. This is a book you will want to read, re-read, and return to again, to fully appreciate the perspective it offers.

This book is a call to take nature seriously, to open our eyes to the simple truth concerning the forces acting upon our self-assembly and the self-assembly of everything that is natural around us, and, importantly, that which is not.

This author does not shy away from important subjects, including pharmaceutical influences on our health or the epidemic that is obesity.

Dr. Galloway's foundational idea of 'similar but not the same' logically describes cell and whole-body symmetry, which is based upon the push-pull forces of biotensegrity. His step-by-step explanations of complex topics, such as cell growth and cell death, make them easy for the reader to digest and comprehend. And, dealing expertly with genetic expression influenced by epigenetics, he allows the reader to appreciate how genes direct cell construction while also calling our attention to the lesser-known hierarchical importance of the role played by Hox genes. And he manages to connect these microscopic realities directly to the big picture of the much broader systems they affect, illuminating our path towards a healthier, better-informed world for us all.

John Sharkey MSc
Clinical Anatomist (BACA, Anatomical Society)
NTC/University of Chester
Dublin, Ireland

Introduction

Look at your hands. Notice the shape, skin, nails, and joints. Now hold them together to see how they are the same. Open them to look at the creases of your palms and the form of your fingers. Starting with the pinkie, see how it measures up against its neighbor. Does the tip of this smallest finger reach to the joint on your ring finger, or past it? And when you compare your left and right hands, notice how they are similar, but not quite the same. For many of us, there will be obvious differences that you may never have noticed before, such as one pinkie longer than the other. As this book explores, learning to look for symmetries, even these obvious patterns that are visible to the naked eye, can reveal a great deal about our bodies – not only about our hands, but also pointing towards likelihoods about our feet, our athletic skill, our bladder function, or even our overall health.

Human bodies are fascinating, but living in a body limits our angle for viewing our own. We can extend the perspective somewhat with simple tools, but even with a mirror, or a camera, our view remains limited. We rarely see ourselves except from the front, most often just our faces. And few of us ever see any part of ourselves but the outside, so studying other human bodies is usually the only way to gain knowledge of what lies inside us. Even while we constantly, intimately experience the sensations our insides give us, we must rely on drawings and descriptions of the inner human anatomy to give us an idea of what structures and inner events produce those sensations.

Da Vinci's Vitruvian Man

Like biologists and doctors who map inner and outer anatomy, artists also study the human body, and it was a favorite subject of the most famously curious mind of the Italian Renaissance, Leonardo da Vinci. He pursued a lifelong study of human form, and his anatomical drawings remain some of the finest examples of their kind. Seeking pattern and symmetry, his artist's eye captured them in the now-iconic series of drawings he called 'Study of the Human Body', depicting a figure that has since become known as 'Vitruvian Man' (after the architect Vitruvius, whose ideas of proportion inspired the drawing).[1]

Standing upright with feet together, the figure fits perfectly inside a square, because the distance across his body, from fingertip to outstretched fingertip, exactly matches his height, from head to heel. Such mathematical proportions abound in this idealized human form. For instance, his height is six feet – six times the length of his actual feet – and, at the same time, exactly four cubits, another ancient unit of measure that cues from the body: the length from the crease of the elbow to the tip of the finger. In the square, the figure's center lies in the pelvis, at the seat of sexual anatomy.

Layering a variation into this most famous iteration of the drawing, Da Vinci animated the figure by spreading out its limbs, with the legs in a wide, confident stance and arms reaching up, all touching the outline of a perfect circle that rests on the baseline of the square. Together, the intersecting

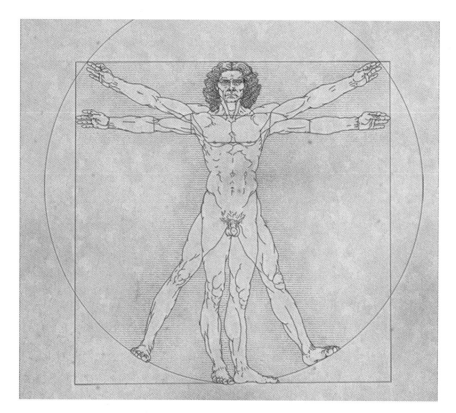

Figure 1

Vitruvian Man by
Leonardo da Vinci.
Redrawn by Paul
Rodecker.

circle and square envelop and define the figure, using visual art as the language to illuminate the profound simplicity of a human body's symmetries. The spread legs also create an equilateral triangle, and the midpoint of the figure in the circle no longer lands at the pelvis but shifts up slightly, to the umbilicus. This stance for the human form is familiar and casual, but the drawing also reflects painstaking organization along multiple dimensions, as the artist has chosen to lay out the figure in a position that balances the figure's biological proportions with mathematical precision.

Vitruvian Man has held lasting appeal across the centuries because Da Vinci's careful representation of human proportion satisfies us – geometrically illustrates something we intuitively know, from the simple observation of our own eyes: that the human body has an elegant balance and formulaic symmetry to it. Seeing these symmetries laid out with precise measurements confirms that intuition.

But while this figure of a Renaissance man presents an idea of perfect proportions, raw scientific data rarely look so neat and tidy. The measurements of actual human bodies range widely, both in size and in relation to one another. In a few countries, we still measure height in feet, a lingering remnant of measurement systems based on natural proportion, but medicine has long abandoned the notion of an 'ideal body'. The rich data of broad population studies provide, instead, averages of actual human size, weight, and proportion, to define the spectrum within which human bodies live.[2] And in this world of statistical analysis, Vitruvian Man may still hold some fascination as art, but he has little to offer medicine. Or does he?

Da Vinci was a scientist as well as an artist, and his multidisciplinary vision gave him a unique ability to observe closely, such that his drawings inform not only the history of art but the history of anatomy. What limited him was not lack of vision but lack of equipment: he was bound by the limits of what he could see with the naked eye. Today we have knowledge not only from external observations but also from centuries of detailed anatomical dissections of corpses. Eventually, surgery also opened the secrets of the body while still alive, and now, finally, noninvasive imaging such as x-ray and MRI reveals the living body – even the moving body – to the ordinary eye, in images that remained hidden to Da Vinci. Thus, in the 21st century, anyone interested in the body can now fill in crucial details unavailable to a Renaissance man, for observing the symmetries between parts and systems. And, as we explore in this book, Da Vinci's method of symmetry-seeking is still relevant and instructive for us today, even and perhaps especially so now that we have the technological tools to fill in those details.

Setting the scale: human-sized

Current technologies stretch our imagination to new dimensions of measuring, as radio telescopes reach out beyond our galaxy while cyclotrons split subatomic specks, and every research venture adds terabytes upon gigabytes to our store of information. For the symmetry-seeking approach to help us manage the resulting information overload, the human body provides a familiar starting and ending place. Protagoras told us that man is the measure of all things, and the human body, male and female, still sets the scale for how we humans see and understand the world. What we can see with the naked eye still circumscribes our daily experience, and knowledge from our extraordinary technologies still must be brought into a viewable human scale if we are to comprehend it. Symmetry-seeking allows us to see the parallels between humans and many other forms and systems, much larger and much smaller than ourselves, as well as our profound effects on those systems and theirs upon us.

As a surgeon, I operate on just such a scale, human body to human body, examining both outsides and the depths of insides. Anesthetized patients grant doctors a direct view of inner structures that the patients never see for themselves. And from one body to the next, the infinite variety of differences between those structures is astonishing, especially in the lower abdomen and pelvis, where as a urologist I work most often. Amidst the many commonalities between bodies, hidden surprises await inside each person's belly. My efforts to understand the reasons for those differences have led me beyond urology into the topics I explore in this book.

From our parents and our ancestors before them, every living person has inherited genetic characteristics that shape our health, across a wide spectrum of variation. All of us notice these variations – those family features – on the outside of the body, while the surgeon also sees the result of those family influences on the inside. And some of those genetically determined characteristics suit modern life better than others. Vitruvian Man represented an imagined ideal of balanced structure, and still does, but even within the range of variations found among reasonably healthy people, we real humans inherit miscellaneous limitations that hamper optimal function in some way or other.

Experience (our nurture) interacts with genetic inheritance (our nature) to encourage or limit the patterns of our development, from our conception forward to any given moment in our lives. Just as conditions of soil, rain, frost, and disease will affect a tomato plant's patterns of growth, what we ingest and the influences of our environment shape our health, including both the factors we choose and those we don't. Very few people have enjoyed truly ideal conditions for development, because most of us have been exposed to unfavorable influences, whether for short or extended intervals – conditions that have affected our health trajectory. Thus, even among the genetically blessed, circumstances of development deliver insults to our bodies and create threats to the fulfillment of our genetic potential for optimal form and function.

Champion athletes tend to have highly symmetrical bodies, with the complete balance of Vitruvian Man. To speed faster than others in a straight line, sprinters or swimmers need equal power for muscular effort on both the right and the left. Many of the rest of us have one leg that is stronger than the other, and we may even have trouble finding pairs of shoes that fit, because one foot is longer or wider. If you are familiar with this slight asymmetry yourself, you may assume that no one's two feet are quite the same size. But some people do have same-sized feet, which are also near-perfectly symmetrical in other ways. Within a family, members tend to follow similar patterns, and our family members' bodies are the ones most likely to be familiar to us, beyond our own, so your knowledge about bodies is likely to be based on that limited sample, which can lead you to erroneous assumptions. Learning about your own personal asymmetries can point you towards other insights about your personal health that few of us know but all of us (and our healthcare providers) could find useful.

Seeking a new viewpoint

Personal limitations from our family genetics are also part of a larger evolutionary heritage. We humans pay a price for the ways we have evolved, trade-offs demanded in compensation for our big brains, especially at the other end of our bodies. If you look at your dog's paws or your cat's, or at the hooves of horses, cattle, or hogs, you won't find notable differences between right and left; their hind feet are mirror images. Funny feet, it turns out, are part of the human condition, not the general mammalian one. To understand why, we have asked Vitruvian Man to assume a new position, with the help of our artist: sitting, with his legs out in a split, arms folded across his chest.

We are accustomed to seeing this familiar figure standing strong, perfect, and invulnerable. But in this unfamiliar pose, the figure reveals a new truth about our human condition. The five centuries between Da Vinci's time and our own are but a moment in human history, not long enough for any profound evolutionary variations to try themselves

out and take hold or fall away, so the human body has not changed significantly since then, and Vitruvian Man is still a fine depiction of human symmetry. What has changed – and is a primary subject of this book – is our knowledge of the inner workings of the body, in particular the nervous system. Viewing a man from the outside, from the front, standing, Da Vinci was like an explorer able to study a familiar shoreline without knowing its uncharted interior. The inner terrain had been contoured by the nervous system, but it was still invisible.

Now that science has revealed how the nervous system works, we can also know that pressing the figure down into the triangle presents the human body in the configuration true to its own development. As in Da Vinci's square and circle, our re-envisioned figure in a triangle retains a halo of safe space around his head, but at the bottom, he now sits directly, uncomfortably on his vulnerable perineum, with no more wiggle room below him. His toes still reach to the margin of the figure, but now his tail end also rests on the very boundary of the shape, his actual, structural bottom on the line. Thus, instead of reaching from head to toe, the figure encapsulated in a triangle reaches from head to tail, like the nervous system. Its spinal cord, which extends nerves outward to all the parts of the body, reaches from the base of the head down to the anus and into a vestigial tail, beyond the level of the spine, at which it extends nerves into the feet and toes (since the legs attach higher up the body).

As we'll also explore, evolution prioritizes heads over tails, a fact evidenced by the size, complexity, and protection of the brain when we compare it to our humble tail end. Slowly, over the millennia of evolutionary time, human heads and brains have become larger than ever, yielding great progress for the intellect, reason, and imagination. But those same forces of selection have created problems for the tail end of our bodies – challenges we can manage better if we understand how to spot the signs, even from the outside, that something may

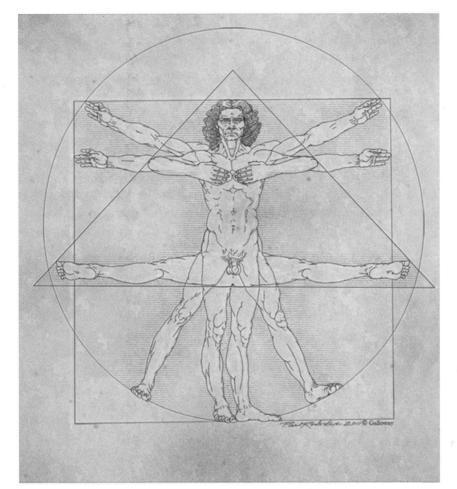

Figure 2

Vitruvian Man revised, with the addition of a triangle. Redrawn by Paul Rodecker.

be amiss. If you have especially funny-looking feet, for instance, you should be prepared for pelvic floor problems, because the nerves that shape the form and function of the feet also shape the perineum, so one will mirror the other in predictable ways.

Nerves and neural development turn out to be critical to health, because some of the balance in growth and form reflects a balanced contribution of nerve density. In each growing embryo, nerves direct the construction project that is the growth of a new human body, laying out the internal order as well as the external, in patterns that mirror one another, if you know what to look for. And although in medical textbooks and discussions of anatomy, the human body is routinely assumed to lack internal symmetry, we shall reveal its astounding, unnoticed symmetrical patterns, by viewing the inner organs as the nervous system does, from the spine. The triangle having opened our understanding to a new view of the human figure, we will play further with our perspective, turning the figure around so that we may view it from the back. This rear view sets the limbs, which reach forward to the front, out of our way, and then all sorts of unexpected symmetries come into view, as we look under the skin to the deep inner anatomy. We will also compare the symmetries between male and female form, as well as between our forms and those of other mammalian species, other animals, and even plants, to deepen our understanding of not only the nervous system but the whole body.

How we got here, and where we are

Since the middle of the last century, an interlocking matrix of massive industries – medical science, agriculture, food, and pharmacy – has come to dominate the factors that most determine human health. In the 1940s, amidst the chaos of global conflict, governments the world over had turned to science, both for better and for worse, to accomplish their wartime ends. After the conflict ended, the war effort left stockpiles of overproduced supplies, fresh scientific discoveries, and excess manufacturing capacity with no market, so postwar commerce turned these products of war to domestic uses: synthetic nitrates from armaments into fertilizer, chemical weapons into pesticides and herbicides, and military rations preserved for troops into prepackaged and processed foods for everyone. At government insistence, agriculture applied new massive-scale industrial methods to boosting food production to feed a hungry world. New agri-science reshaped agribusiness so thoroughly that a handful of ultra-high-yield crops such as corn and soybeans transformed not only the physical landscape of farms but also the marketplace of food products and, ultimately, our national health and wellness – and not, on balance, for the better.

Thus, after more than half a century of industry and government together pushing production of certain foods, the inescapable reality is that 'eating is making us sick', according to physicians such as David Kessler, after his studies of American eating habits and government food policies, including *Your Food is Fooling You*.[3] A wealth of research and a number of excellent books, such as Colin Campbell's *China Study*,[4] have examined the relationship between diet and health risks, including today's widespread incidence of heart disease, diabetes, and cancer. Overall, science presents compelling evidence that good nutrition is the true foundation of health.

As Michael Pollan explored in *The Omnivore's Dilemma*, however, we now face a daily difficulty about what to eat.[5] As a physician, I can confirm what you may already realize: that this question is one of life and death, because most of our major life-threatening diseases – those that threaten longevity, quality of life, or both – are the result of daily personal choices. And the most important of those choices, above all, concerns what we eat.

Meanwhile, instead of the mindset and methods of the pastoral farmer, who grows food by nurturing the soil and tending to livestock, monoculture and industrial agribusiness prevail, and the idea and methods of war linger. Medicine's war on cancer drags on, and we declare war on any stubborn peacetime challenge that begs for reform and improvement: on poverty, on terror, on drugs. These 'wars' also distract national attention from even more widespread hazards, some of which have well-defined paths towards solutions. Amidst all the attention given to illegal and street drugs, for instance, we have been slow to notice the threats from pharmaceuticals in general, even though adverse events and unintended effects of medical treatments now rank third among the leading causes of death in the US. We have record medical spending but worse health outcomes, and life expectancy is now projected to be shorter for our children than for their parents, as obesity swells and diabetes, heart disease, and cancer flourish. The progeny of 1940s weapons of war, redirected for domestic use, now threaten our own human bodies, far more than any foreign foe.

Symmetry-seeking as a tool

Rather than lacking information on diet and health, we are drowning in a deluge of medical data and an overwhelming mass of details. Various books, as well as various experts whom we trust, give us conflicting guidelines for how to make good decisions about our health and, especially, our food, to the point that more learning sometimes bogs us deeper in confusion, rather than relieving the worry that we are making the wrong choices. We flail in the flood of information from countless sources and points of view, with contradictory opinions. In our work, our entertainments, and our daily round, signs and sounds and pleas and advertisements pound us with grabs for our attention: buy this, be afraid of that, try this new product. Many of these appeals and promotions relate to health, and the volume of the outpouring creates confusion and anxiety for anyone trying to process it all, day-to-day.

I am particularly interested in these questions of overall health now that I have practiced medicine for decades, as I see my patients suffering under the pressure of the flood. And the overload of information, including uninvited promotional material, is even worse for doctors and medical researchers. In the current explosion of data collection, publications and medical journals pummel us with much more information than any one doctor could ever read. Medicine's answer to overabundant data is usually hyper-specialization: learning more and more about less and less, more detail about an ever-shrinking area of the body, while forgetting what we may have once learned, years ago in anatomy class, about the body as a whole. But unlike the medical specialists who dispense advice, the patient (or a healthy human being seeking to avoid becoming a patient) doesn't have the luxury of giving attention to only one system in her body. She experiences her body as a whole, each part affecting the others and the health of her body also affecting her state of mind, which in turn continually shapes her health.

Fortunately, that same science which has inundated us with data also lights the path out of the mire, by giving us the tools of symmetry, whose principles are intimately intertwined with those of evolution. In the late 1800s, while Darwin was introducing his theories of natural selection that comprise the more widely familiar aspects of evolution (he published *On the Origin of Species* in 1859),[6] Scottish biologist D'Arcy Thompson was discovering the physical principles of economy and transformation that shape both living and inorganic structures, articulated in his *On Growth and Form* in 1917.[7] Long before science had discovered genes and biochemical mechanisms, he already understood that geometry was at the heart of shaping both shells and horns, leaves and flowers, and even teeth and bones.

Science has since discovered how and why simple geometry plays such a major role in shaping biological complexity and just how chemical gradients interact to generate patterns in biology, insights Alan Turing first foresaw in 1952.[8] Contemporary researchers have pieced the puzzle together. Biochemist Sean Carroll articulates the scientific discoveries in evolution and developmental biology that have revealed the robust molecular mechanisms for growth and form, in principles that provide the basis not only for animal forms but also for human variation and diversity.[9] Psychiatrist and neuroscientist Iain McGilchrist, in *The Master and his Emissary*,[10] has described the contemporary human brain, its symmetry and asymmetry, going so far as to suggest the role of asymmetry in shaping the differences between eastern and western worlds.

Their writings confirm the underpinnings of what I've discovered from my clinical experience and write about in this book: that learning to look for symmetries can reveal a great deal about our bodies - their functions, vulnerabilities, best treatments, and most promising routes to staying healthy. And my research has ultimately brought me back into human origins - embryology - and into relevant topics in neurology, a longtime research interest for me. It has also carried me beyond the human body into the many environmental, institutional, and behavioral factors that affect our health and are affected by the choices we humans are making, especially our choices about what we eat.

In my work as a physician and surgeon, I work closely with specialists in fields other than my own - urology - such as gynecology, gastroenterology, pediatrics, and neurology. And as you'll see in the coming chapters, my endeavors to understand a bigger picture have led me to examine the work and writings of experts far afield of my own backyard, not only within medicine, the healthcare professions, and public health but reaching as broadly as the farming industry and the grocery store business, or looking at history and mathematics. The book itself has been the result of several years of focused conversation around this wide range of topics with my co-author, whose specialties lie in language, technology, and composition – discussions all centered around and returning to the idea of symmetry, pattern, and the big picture.

The book is intended to open a dialogue and to define new questions. It is our hope that both individuals who want to learn more about the human body and experts from a wide range of disciplines will find cues here to send their questioning on new trajectories.

We hope, too, that you will gain a new perspective on how to make best use of experts. In a hyper-specialized world amidst an overload of data, decision-making is a challenge. It helps to understand what expertise you yourself bring to the table, what other experts offer and where their blind spots likely lie, and how to merge what you know with what they know, when a decision is important. When overwhelmed with information, it's tempting either to abdicate decision-making to ostensible experts, or, since they and their biases have failed us before, just to ignore them and go it on our own. With the help of symmetry-seeking, we offer a middle way, and a call to expand your view and look into other fields, other expertise, other silos, in the confidence of your own perspective, all while understanding where your own expertise does and does not lie.

My ventures here into other fields offer an example of that very sort of action. Cowed by others' deep specialization, we have become afraid to look into other fields even when they affect us mightily. But it is not audacious – rather, vitally necessary that we muster the courage to look beyond our own silos and consider how others' learning informs our own. Modeling the task, there are chapters here on hormones, the pharmaceutical industry, plant farming and the problem of ammonium nitrate fertilizer, and even a brief sortie into psychology and the brain-body connection. Experts in any of those disciplines can (and no doubt will) correct any missteps and hone my account, which I welcome. If we are to make the most of the extraordinary knowledge that the 21st century offers us, we mustn't be afraid to look at and seriously consider work outside our own ever-smaller, if deeper, domains.

This book's primary application of our symmetry-seeking approach is to questions of human health. As you consider what your own areas of expertise may be, one is certain, and of supreme importance: you are the unparalleled expert on your own body. We hope this book deepens your expertise.

The wonder of likeness

Difference is easy to notice, while similarity is often hidden or, sometimes, so obvious and familiar that it is easy to ignore. But the quintessence of things lies in their many similarities, much more than in their differences. Symmetry-seeking therefore asks that we search for similarities instead of differences, or for likeness within difference – and that we look for certain types of pattern and repetition – even when disorder and randomness are more evident. Symmetry-seeking is a deliberate practice of looking for similarity.

Variety usually captures our attention and stands out from the same-old same-old. It's not for nothing that the 'variety show' is a longtime standard in entertainment or that the novel is called 'the novel'. We love to discover the new, the unfamiliar, the unexpected. Doctors, too, delight in encountering the unusual patient and are fascinated with the rare diagnosis – we will look at a few such extraordinary medical cases in this book, too, as exceptions that help us understand a rule.

But scientists know that while the rare exception can be fascinating, it is predictable repetition and sameness that reveal the profound simplicity of nature, which can be even more intriguing. When we discover that one thing is strikingly like another, when there are patterns within patterns that we didn't expect to find, these are the foundational revelations of science. When we find, furthermore, that a pattern over here is unexpectedly just like that one over there, widely different and seemingly unrelated, yet strikingly similar in ways that change how we need to deal with both, then we've made a truly useful connection.

Certainly in medical diagnosis, a doctor is looking for patterns of cause and effect – an inquiry often begun already by patients themselves before coming to the clinic: 'it is worse in the morning', he might report, or 'it only hurts when I cough'. And a doctor or researcher looks for patterns across patients, observations that are the essence of the expertise which informs diagnosis and directs effective treatment. In any study of a given health intervention or pharmaceutical trial, researchers are looking for patterns: when some patients try this or that treatment and others do not, can we find a significant and reproducible difference in the pattern of survival or recovery? Within any scientific experiment, the researcher is usually looking for a robust outcome, and the ability to replicate the results is a fundamental measure of its reliability; to be trustworthy, the pattern should reappear faithfully again and again in subsequent experiments.

Predictable sameness has a compelling simplicity, an elegance which can drive not only scientists but mathematicians or any theorist to seek the grail of a grand unifying theory. While this book does not pretend to lay out such a theory, we do share that urge towards compelling simplicity. We seek out and explore the underlying, previously unnoticed patterns of amazing sameness that can change how we see the world, how we sort through information and, more particularly, how we can apply this perspective – of symmetry-seeking – to make decisions about our own personal health.

Medical science has amassed mountains of health-related information without any clear paths for ascent and few directions to navigate the range, leaving us intimidated by the scope of the climb or lost in the clouds of overabundant data, and hyperspecialization is both a cause and an effect of the challenge. The symmetry-seeking principles we propose here offer a path, a methodology, to help guide us as we sort through some of the detritus in the search for useful clinical data. After I noticed unexpected similarities between internal organs – patterns of repetition in both structure and function between the kidneys and the lungs, for instance – my exploration of these symmetries carried me out to looking at other symmetries in the human body, such as the way foot problems tend to correlate with bladder problems. These symmetries then began to inform my diagnosis and treatment of patients as well as my search for the reasons behind these correlations.

This book seeks to share the principles of symmetry-seeking that have emerged and to demonstrate their use by showing them in action. Most of the scenarios we present here involve decisions related to human health.

These symmetry-seeking principles also form a methodology that any researcher can apply to topics beyond questions of health, and in fact one of the central themes of the book is the interconnectedness of the human body with everything outside it, both animate and inanimate.

We offer the perspective of symmetry-seeking as a way to explore simple patterns beneath the surface, repetitions that shape the complex landscape of human health and disease. You won't always find symmetry where you expect it, but even its absence can be crucially instructive, and its presence, often in unexpected places, can surprise and inform your knowledge. Most important, the search itself gives us a frame of reference on which to hang the new information we constantly encounter in our learning, to help us decide what matters most for our health – and, for health care providers, to guide our patients well.

Instead of trying to ground your health in a list of ever-shifting food group categories, you will learn some principles for asking your own questions about what is good or bad for you. A wider view of the food chain reveals each link coupled in turn to the last, each signaling system shared with others. In such a chain, then, growth hormones for livestock might drive human obesity, and long-acting drugs to promote milk yield might spill over to drive precocious breast development in children.

We invite you to go beyond the familiar outer surface of our human bodies to explore deep inside, seeking symmetry. Expect to find forms, both internal and external, that are made to a pattern: surprisingly similar, but not quite the same, symmetrical but not identical. Because all living structures start small – not just small, but as one tiny microscopic single cell – every individual of every species begins as an infinitesimal packet of instructions to be used over and over again. With stunning economy, nature packs all its know-how into this cell, using the same laws of physics and chemistry, the same tools for growth and development, the same constraints, within each organism and across species. So we are not made of different as much as we are made of same. In this book, we will ask you to step back as far as you can from the details, to see them within the context of the big picture, to look for patterns in the whole, of which humans are a part.

Some principles of symmetry-seeking

Looking for similarity is the core principle of symmetry-seeking. All other ideas in this approach rest on and further this one practice, of looking for certain kinds of repeated patterns. A search for pattern, generally speaking, is central to many forms of inquiry. Our methodology of symmetry-seeking focuses on questions of balance and wholeness, especially by looking at similarities that are structural, functional, and developmental.

To search not just for similarity but for symmetry-based similarities within a thing or between two entities – whether in an organism, a system, or any other object – you must start by choosing an axis of symmetry. Drawing this line of comparison allows you to see structural and functional repetitions that you may have not noticed before. Placing the line in unexpected locations as well as the obvious

ones (such as down the front of the human body) or between entities that have no obvious similarities can help to illuminate hidden symmetries.

To reap the rewards of symmetry-seeking, it is essential to remember that many of the patterns you discover, while notably similar, will not be precisely the same. If you insist on exactitude in the patterns, you won't recognize them as patterns, and you will miss the insights that symmetry has to offer. This softness around the outlines of these similarities, the inexactitude of the symmetry, simply confirms that there is a range of natural variation across living systems. Our method requires that we approach the looking with a new lens, not only a wider view but a softer focus, to be able to see the patterns.

Here are some essential aspects of the symmetry-seeking habit that we know thus far, which you will see recur throughout the book, as we dive into explorations of the human body, health, and the systems that affect the body and the planet:

- **The big picture**. Symmetry-seeking insists that we step back from details to see them in the context of a broader view, such as in the reflections between feet and pelvic floor functions.

- **The whole, not just parts**. When the broader view includes a whole – whether an organism or another kind of system – the search for symmetry asks that we look at the whole object and see its parts in the context of the whole.

- **Perspective**. In placing the axis of symmetry, it is important to adventure forth from the usual angle to find new viewpoints, such as from the back, the top, or the side.

- **Scale**. Some of the most revealing similarities occur between entities of widely varying size. The search for symmetry often requires scaling one of them larger or smaller to facilitate comparison across the axis line, such as when comparing a complete human to a single-celled bacterium, to better understand DNA.

- **Function**. Not only the structure of organs, organisms, or systems but also their functions can reveal symmetry. Thyroid and ovary, for instance, share not only similar structures but similar functions, both regulating growth and metabolism, biorhythms and balance.

- **Adaptive purpose**. A structure and its function evolve in a given environment to serve some purpose. This analysis works in both directions, because assessing purpose can reveal symmetries with similar structures, and similar structures usually point to symmetry of purpose.

- **Developmental simultaneity**. A hallmark of this idea of symmetry-seeking is attention to developmental pattern – noticing similarities between structures that appear at the same moment in early development, such as in the human embryo. The larynx and the urethra are such a pair, both gateways that open and close to separate the internal spaces from external, sharing similar structure, function, and surprising links between events in one causing parallel occurrences in the other.

- **The long view**. Where developmental patterns ask us to look back from the present moment to see how a system originated, symmetry-seeking also insists that we take the long view forward. What happens to that same system later, as it ages, and what does symmetry reveal and predict about how it will function or fail? What happens downstream, as one system interacts with others? This ability to take the long view also gives symmetry-seeking some predictive power, such as the way we now know fetal exposure to certain chemical agents will predict higher cancer risk later in life.

- **Balance and compensation**. Nature strives relentlessly for balance, so if some factor creates an imbalance, a system usually adjusts somehow to compensate, to counterbalance the asymmetry. Learning to spot patterns helps us to see when they are broken – a clue to some significant shift in a system's growth, health, or function. And any notable disproportion in a system – an element that stands out or overwhelms other elements – is an imbalance that, through the lens of symmetry-seeking, sets us on a search for factors that compensate for the asymmetry, such as when loss of sight brings greater acuity of hearing. Mismatch always seems to work its way back to matching, to a rebalancing, whether over a broader span of space or a longer period of time.

Ultimately, it is our hope that, in addition to guiding you towards a new and practical means of making decisions in the age of information overload, our sharing these ideas will help to open a conversation about the nature of this approach, this methodology, for researchers who wish to put it to systematic use. The chapters demonstrate symmetry-seeking in practice, applying it to understanding the human body, to demystifying some aspects of medicine, and to looking at industrial systems that most directly affect human health: big food, big pharma, and agribusiness. Throughout, you will see the patterns of thinking outlined here, as we model how symmetry-seeking reveals mismatch and helps us recognize the heavy challenges those systems currently pose to our health.

References

1. Isaacson W (2017). The inspiration behind Leonardo da Vinci's Vitruvian Man. Online: https://medium.com/s/leonardo-da-vinci/the-inspiration-behind-leonardo-da-vincis-vitruvian-man-974c525495ec.
2. Stampler L (2014). How the average American man's body compares to others around the world. Online: http://time.com/3551742/average-american-man-body-comparison-photos/.

3. Kessler DA (2012). Your Food Is Fooling You: How Your Brain Is Hijacked by Sugar, Fat, and Salt. Roaring Brook Press.

4. Campbell TC (2006). The China Study: The Most Comprehensive Study of Nutrition Ever Conducted and the Startling Implications for Diet, Weight Loss, and Long-Term Health. BenBella Books.

5. Pollan M (2011). The Omnivore's Dilemma: The Search for a Perfect Meal in a Fast-Food World. Bloomsbury Paperbacks.

6. Darwin C (1859). On the Origin of Species by Means of Natural Selection, or the Preservation of Favoured Races in the Struggle for Life. London: John Murray.

7. Thompson DW (1917). On Growth and Form. Cambridge: Cambridge University Press.

8. Turing AM (1952). The chemical basis of morphogenesis. *Philosophical Transactions of the Royal Society of London. Series B, Biological Sciences*, 237(641): 37–72. Retrieved from http://www.jstor.org/stable/92463.

9. Carroll SB, Grenier JK, Weatherbee SD (2004). From DNA to Diversity: Molecular Genetics and the Evolution of Animal Design. Oxford: Blackwell Publishing.

10. McGilchrist I (2012). The Master and His Emissary: The Divided Brain and the Making of the Western World. Yale University Press.

Human architecture

One hundred years ago, the famous Scottish biologist D'Arcy Thompson spotted the physics of living things: 'Cell and tissue, shell and bone, leaf and flower are so many portions of matter, and it is in obedience to the laws of physics that their particles have been moved, moulded and conformed'. A great observer of nature, Thompson was intrigued at how often 'the forms of living things and the parts of living things can be explained by physical considerations and to realize that in general no organic forms exist save such as are in conformity with physical and mathematical laws'.[1] Our human bodies are one of those living things, also 'so many portions of matter', and our living particles have been moved, molded, and conformed in accordance with those same general physical laws. In his recent book *Design in Nature*, Adrian Bejan arrives at similar conclusions about an even broader range of organisms and systems. Exploring familiar forms such as trees and rivers, vascular branching, and respiratory airways as well as electricity in circuit boards, he proposes that the shape shared among all these forms arises from their common purpose – that they are all systems evolved to promote ever more efficient forward flow.[2]

What Bejan calls 'flow' is a shared feature of many natural systems, from biology to geography, all of which involve movement. In a storm, thunder and lightning reflect the sound and sight of electrical discharge flowing from high-energy clouds to less-charged clouds. Flow always favors the path of least resistance, so whatever needs to move takes the easiest way it can find. Electrical charge hops within and between clouds, building potential for discharge until, in an instant, one big streak bolts to the nearest high ground, cascading down to strike a tree or rooftop. The forked lightening is instantaneous and fleeting, but its pattern mirrors that of other familiar forms, whether tree roots, river estuaries, or small veins that come together to form larger ones (venous arcades). The pattern repeats again and again: many and smaller unite to become fewer and larger and, in the process, weaker and slower become stronger and faster.

What is flowing doesn't matter nearly so much as the fact of flow, according to Bejan, who tells us that these naturally made fractal structures exhibit shared shape and form because of their shared function and purpose, always promoting forward movement. The physical laws that govern the patterns of flow will apply equally to all that is flowing, and the field will be shaped according to those shared efficiencies and constraints. Whether electrical charge, water, blood, or air, if the system is about flow, the physical laws will define and channel that flow in familiar and recurring patterns, which reflect optimal streaming.

Just as Bejan makes the case for patterns of flow, so it is for patterns of growth. When D'Arcy Thompson observed the striking patterns of similarity between many animals' shells and bones, he posited that they shared common features of form because they shared common laws of growth. Those laws, in turn, emerged from parallel needs for function, which create similar physical constraints and efficiencies.

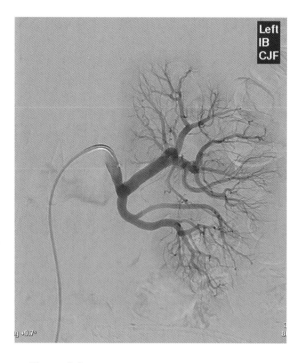

Figure 1.1

Selective renal arteriography – left kidney.
This X-ray image of the kidney is obtained by
injecting contrast medium into the renal artery
to display the vessels that carry blood from
the heart to the kidney tissue. The vascular
bed has the familiar form of a tree, with one
or more trunks, dividing into a few major
branches, more smaller branches and many
tiny twigs. Another tree of vessels collects the
blood from the tissues as tiny vessels called
venules that join together to form larger veins,
and in turn, large venous trunks that drain the
blood back to the heart.

Growing in organisms – like flowing, in the systems Bejan examined – is governed by fixed natural laws. Those rules of growing and flowing all generate similar shapes, which are therefore widely familiar to us all. We shall explore some of what is now known about the rules of growth and discover that, in living systems, not only the rules but also the ancient tools for growth are shared and common to all.

The surgeon must often raise a flap of adjacent skin and turn it to a new position, just as the tailor might cut cloth and sew it in place to cover a contiguous part. Easier for the tailor, because fabric needs no blood supply; skin, on the other hand, must be nourished. It depends on neighboring areas of skin, which must be the bridge to adjacent blood vessels that can keep the skin tissue alive. In surgery, experience teaches us that skin flaps depend critically on the base of skin they attach to, and the geometry is simple: to survive, the skin flap must be no longer than its base is wide. The wider the base, the larger the skin flap can be. But the flap must never be longer than the width of its base, else some of the skin at the tip will perish and die. The foundation is key. The ideal flap is square-shaped, attached along one side but free on the other three edges. And for living skin to survive, the skin flap must never be longer than it is wide, so only a square-shaped flap provides the ideal ratio – maximum area for coverage but without risk of skin breakdown.

These flaps of living skin that we sew together during surgery follow a predictable model, on a principle that applies not just for repairs in adult patients but, viewed through the eyes of symmetry-seeking, for growth during early life, whether in an embryo, fetus, or child. Whereas the skin is flat, the structures the embryo is growing are fully three-dimensional, but the mathematical principle is the same: the dimensions of the base will support only a given amount of living tissue and no more. So, just as living skin needs to attach to a base of a length no larger than itself, to survive, newly forming body parts and vital organs each need a base with a larger volume than themselves as a foundation of support from which to grow. We grow, then, from a base of big parts out to smaller parts, which comes as no surprise: we grow our legs, then our feet, and then our toes. A careful look at the patterns of that growth, in keeping with Thompson's findings, does reveal a surprise, however: a constantly predictable set of proportions in those dimensions.

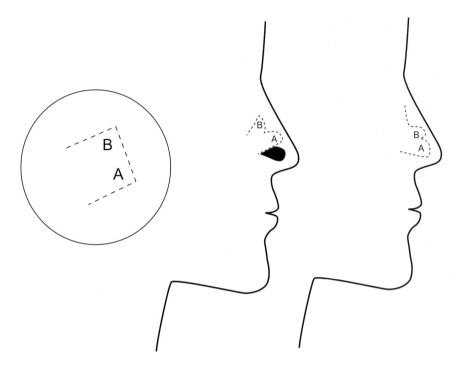

Figure 1.2

Skin flap. In raising a flap of living skin, the length must not exceed the width of the base or the tip of the flap will die.

Forearm, hand, and finger proportions: the golden ratio

When you look at your open hand, you can compare the distance from the tip of the middle finger to its base where the finger joins the palm. You will see that the ratio of lengths, tip of the finger to the distal crease (behind the knuckle nearest the tip), and distal crease to the base, is the same as the ratio of the whole finger to the length of the palm, measured to the crease at the wrist.

The length of the whole hand is, in turn, the same ratio to the length of the forearm, from the crease of the wrist to the crease of the elbow. Though we are all different sizes, some shorter and others longer, making absolute lengths less useful, the proportions are reliably consistent, like Da Vinci's Vitruvian Man. You can also see that the part nearer the base of a limb is always a little thicker and more substantial, because it is the foundation upon which the next section is grounded. Like a cone, the wider base supports the lesser, more distal segments.

When we compare two numbers, such as these measurements, we call that relationship a ratio.

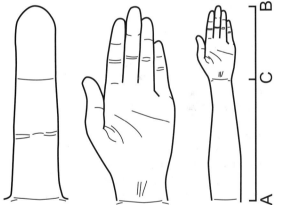

Figure 1.3

Golden ratio proportions, finger to hand and hand to forearm.

But when we compare two ratios, we call that relationship a proportion, such as a part of the finger is to the whole as the whole finger is to palm, or finger is to palm as whole hand is to forearm. And this proportion we see in our limbs and digits is called 'golden', for its qualities are not only exquisite and

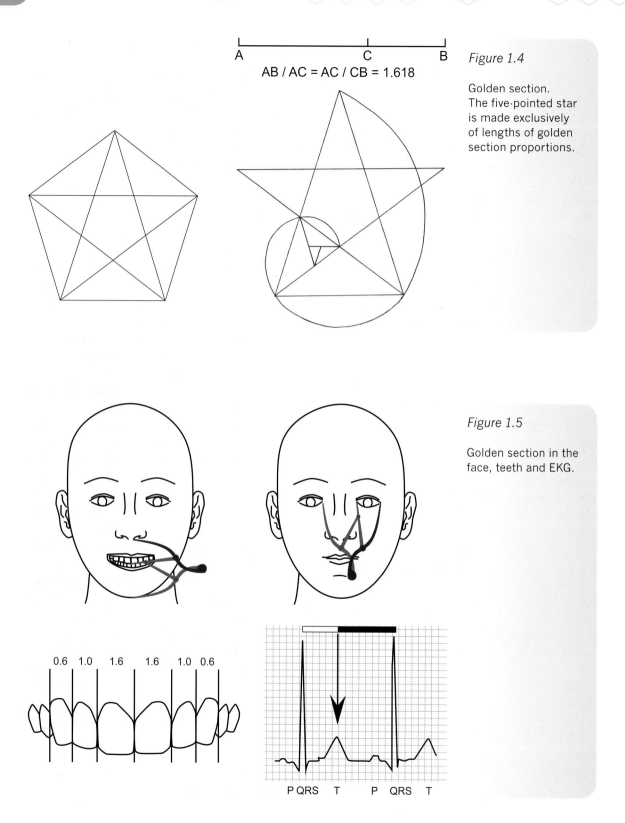

AB / AC = AC / CB = 1.618

Figure 1.4

Golden section.
The five-pointed star
is made exclusively
of lengths of golden
section proportions.

Figure 1.5

Golden section in the
face, teeth and EKG.

0.6 1.0 1.6 1.6 1.0 0.6

P QRS T   P QRS T

astonishing but also unique and, at the same time, universally present – not only in our human bodies but throughout the living world. The 'golden section' divides a line such that, through ever-larger body parts, the length of the shorter portion is to the length of the longer as the longer part is to the whole.

Another way to define the golden section would be the division of a line into two segments such that the whole line is to the longer length as the longer is to the shorter. This sounds so simple, but it is also profound. The golden section could equally be called the growth section, because this proportion is the hallmark of living systems, even our own facial proportions. Look at your face in a mirror and seek out the golden section. It is in the form of your face and the size of your teeth.

We are all different in detail, but broadly similar in form. Human faces evolved to be similar enough to other humans to be able to tell humans from

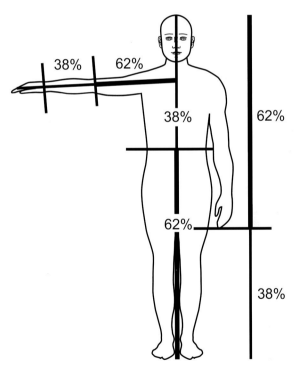

Figure 1.6

The umbilicus in this figure marks the golden section.

beasts but different enough to tell one human from another. Hundreds of millions of faces on passport photographs and each one just a little different from the next: like nature's specimens, endless variations on modest themes. And not just faces: look at your torso or your own standing figure, simultaneously yours alone, but also like the rest. And somehow, we are all made in golden section proportions.

Tensegrity

Relatively little is known about the forces that guide molecules to self-assemble for creating proteins, cells, and tissues, but it is clear that all living structures are constructed using the same basic architecture. Just as the human body is a balance made of compression-bearing bones and flexible or tension-bearing muscles, tendons, and fascia, so all living systems at all levels of scale are made the same way: stiffer struts balance with stretchy strings that hold our structures together, even at the level of our individual proteins and single cells.

In large-scale construction projects, engineers have found ways to design lightweight but sturdy structures that mimic nature, seeming to hold themselves together in midair as if by magic. This kind of architecture, called tensegrity, allows complementary opposite forces to equilibrate throughout the whole and stabilize the shape and form. We see this pattern in some very modern buildings and bridges, which rely on the strangely simple structure of like parts linked together in set proportions with others to counterbalance them.

The term 'tensegrity' was first coined by architect and inventor Richard Buckminster Fuller, combining the qualities of 'tension' and 'integrity' to describe structures that comprise a balance of tensioning and compressive forces distributed and balanced within a structure. Always acting together as complementary opposites, tension elements are arranged in fixed proportion with compression elements. All the structural components are already prestressed in tension or compression, and mechanical stability and strength come from the way the entire structure distributes and balances force and energy.

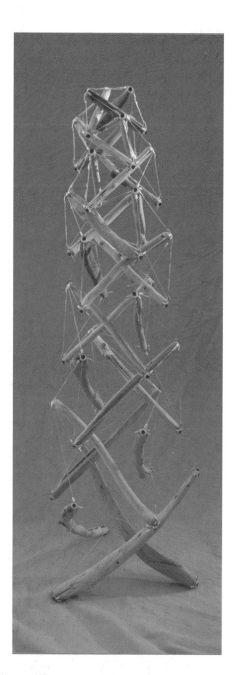

Figure 1.7

Tensegrity architecture: a driftwood tower constructed by Bruce Hamilton.

gravity, but tensegrity structures distribute tension continuously throughout the whole. In tensegrity architecture, increasing tension in one place simultaneously alters tension throughout the entire structure, even in elements on the opposite side. In this way, the construction maintains structural integrity by balancing the global increase in tension with a symmetrical increase in compression, in certain other members spaced throughout the structure.

In bridges or domes, tensegrity structures bring a maximum amount of strength for the least mass of building material. This same economy, and its consequent adaptability and flexibility, also makes tensegrity architecture the perfect choice for building living systems, where it is called biotensegrity. This concept was first introduced by orthopedic surgeon, Dr Stephen Levin, who has pioneered the ideas of tensegrity architecture and led the way for others to explore these ideas in biology and medicine. This method of construction scales from the macroscopic level – the whole body and larger – down to the microscopic and smaller: cells and molecules, including proteins and DNA. All these systems within systems stabilize themselves through the principles of tensegrity architecture.

In his recent book *Biotensegrity: The Structural Basis of Life*, Graham Scarr describes this living architecture.[3] This architecture applies at all size scales, progressing from atoms to the basic building blocks of protein molecules – including the regulatory bases of DNA – and on to individual cells, tissues, organs, and whole bodies. At the molecular level, electrical charges provide the pull and push of attracting and repelling, manifesting the physical laws that shape protein, starch, and lipid molecules. In cell walls, complex chains of fatty acids align like mini-magnets according to physical laws of the hydrophilic and hydrophobic, self-assembling cell-wall structure in ways similar to soap and water making bubbles, or ice and water vapor making snowflakes.

Cell structure

Ideas about cell structure have changed over the years, as more has been discovered about cellular

Traditional man-made walls or bridges gain their stability from the continuous compression of stacking one on one, relying on the constant pull of and subcellular architecture. When I was at medical school, doctors imagined the human cell to be like a tiny balloon filled with jelly, its outer membrane like a little bag that enclosed its contents. Cells were known to have an internal framework, or cytoskeleton, composed of microfilaments and microtubules, but their functions were very poorly understood. Another mystery was how single cells of the body moved or might behave when placed on different surfaces. In a cell culture they would usually spread out and flatten on a glass surface or in the plastic culture dish.

In 1980, Albert K. Harris demonstrated that when grown on a thin, flexible sheet of silicon rubber, cells would contract and become more spherical, even puckering up the silicon into folds. Harris showed that one kind of cell, myofibroblasts – the cells involved in tissue healing and repair – could behave like mini-winches, laying down strings of collagen then exerting remarkable traction forces; in living tissues myofibroblasts are found in wounds, where their function is to pull the wound edges together and help with healing. Cells are dynamic, not inert, and their surfaces are covered with attachment sites now known as integrins. These and other adhesion sites and receptors physically connect a cell to its anchoring substrate and to its neighbors. In addition to attachment, they are responsible for cell-to-cell signaling and information exchange.

In our updated understanding of the cell, it is still surrounded by membranes and filled with viscous fluid, but now we also understand that it has a distinct architecture of microtubules (its stiffer struts) and microfilaments (its stretchy cables), all engineered to provide shape and resilience, to stabilize the cell structure and mechanics. Cells derive support not only from their cytoskeleton of microtubules and microfilaments, but also from the extracellular matrix, the anchoring scaffolding to which most cells are naturally secured in the body.

A cell nucleus is like the yolk of an egg, nested inside the rest of the cell the way a yolk is settled inside the egg white. The nucleus – this sac within a sac – contains all of a cell's DNA, the genetic coding that tells an organism how to grow. Both the surface of the whole cell and the surface of the nucleus inside are constructed with the very same balanced architecture of membranes, microtubules, and microfilaments, and both are susceptible not only to push-and-pull forces at work within the cell but also to the transmitted pushing and pulling at the cell surface. The surface has its own responses to signals from forces outside the cell, forces that

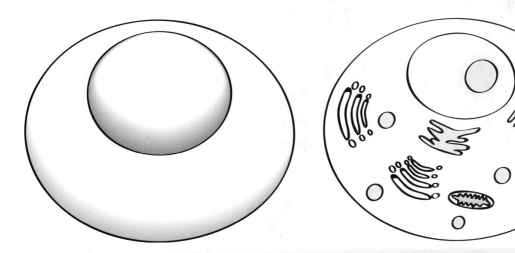

Figure 1.8

Hen's egg and a typical animal cell.

modify the cell's contours and the behavior of the cell, prompting the nucleus to switch cell activity between different genetic programs.

When cells' cytoskeletons stretch to the point of becoming long and drawn out, their shape is signaling that they need more surface area, as in wound repair, to cover the surrounding substrate – a demand which in cells requires division. Push-pull forces trigger the system that directs the cell to divide as needed into two, or to differentiate into another type of cell. But if messaging reflects the opposite – that there has been excessive growth, surplus to requirements, and the cells are overcrowded – then the genetic program can instead redirect toward telling more cells to die off than are currently doing so, in what is called programmed cell death (a process that plays an important role in healthy human development, as Chapter 2 will explain).

Biotensegrity architecture is an answer to cellular requirements and demands for growth. This remarkable structure offers stability, strength, economy, and signaling all at once. And this method of building scales to the whole body, because the needs of tissues, organs, and all the other parts throughout the human body are largely the same as the needs of the individual cell. By arranging their cells using biotensegrity architecture, living systems can use continuous tension and local compression to conserve energy and reduce mass. Biotensegrity is everywhere in nature, because it works at all different scales and sizes, whether within the protein building blocks, the cytoskeleton, cells, tissues, or complete living systems.

Model-making has proved to be a most valuable way of exploring tensegrity architecture, and when one builds and examines the models, their geometry reveals the omnipresent 'golden section' proportions like those in our fingers, hands, and arms. We need only simple geometry to explore tensegrity, whose wondrous simplicity lies in a remarkably consistent mathematical relationship. It is worth a look at some of nature's favorite mathematical proportions to understand the extraordinary sturdiness and consistency of symmetry's foundations. Knowing the basics of these relationships will also allow you to be on the lookout for these quantitative patterns

of proportion in all manner of systems, as you apply the symmetry-seeking way of seeing.

Nature's numbers

In 1975, a researcher at IBM, Benoit Mandelbrot, coined the term 'fractal' to describe the intricate-looking, eerie beauty of endless curves produced using computing power as a random number generator to propagate patterns from a mathematical formula. He built on the work of French mathematician Gaston Julia, who was the first to describe these forms by exploring the mathematics of shapes generated by formulae – 'Julia Sets' – that included an imaginary number, the square root of minus one. These self-similar computer-generated shapes and figures had extraordinary features and appeared to be not two-dimensional and not quite three-dimensional, but somewhere in between. Mandelbrot's equations, known as the Mandelbrot Set, generate forms that constantly grow and feed back on themselves, eventually returning back to the original. Like the natural world, these computer-generated fractal forms exhibit extraordinary complexity, but within themes of recurrence and overall simplicity.

The rational numbers that we learn in basic mathematics can only define and measure fixed forms, which may be beautiful but will always be lifeless – whether triangles, squares, and circles or three-dimensional blocks, columns, spheres, and crystals. Most of the everyday functions of numbers – for schoolwork and accounting as well as all kinds of measuring, calculating, engineering, computing, and other aspects of business and trade – depend on rational numbers.

For living organisms, on the other hand, nature uses fractal geometry – amazing shapes that depend on irrational numbers. Life means movement, and nature's formula for creating living things generates a self-replicating pattern that can repeat and repeat endlessly, at all levels of magnification. The resulting shapes are irregular compared to a straightforward circle or square, but highly regular – amazingly so – in their self-similarity at all levels of their scale, from the micro to the macro, and from the atomic to the galactic.

If you studied a little (or more) higher math, you may remember that the numbers called 'irrational' are so named not because they are mad but because they cannot be expressed as a ratio – a fraction – such as 3/4, 8/5, or 59/1 (because ordinary whole numbers such as 59 are among the rationals). When the fractions are written as decimals, the defining characteristic of a rational number is that the decimal expansion either stops (as in 0.25 or 1.7) or it eventually repeats itself (such as 0.3333.., the decimal representation of 1/3, or 0.0909.., which equals 1/11). As a measurement for a physical form, then, a dimension that can be expressed as a rational number is closed and tidy. It has an end.

In stark contrast, the open growth spirals and branching tree forms that define flow and living shapes are the product of fractal geometry and irrational numbers. Everywhere throughout our natural world, these lively curves and beautiful fractal forms are generated from the mathematics of irrational numbers. These forms replicate the contours of clouds and coastlines, ferns and trees, and natural patterns of flow and turbulence. As György Doczi, an architect interested in these natural forms, explained: 'Irrational numbers are not unreasonable; they are beyond reason, in the sense that they are beyond the grasp of whole numbers'.[4] Rather than the closed perfection of a circle, nature prefers the open-ended movement of a spiral, even within what seems to be a circle. A flower, then, sports a spiral of petals, and a shell a spiral of chambers within chambers. Instead of an exactly repeating pattern, an irrational number creates forms that build out from themselves by replicating but continually scaling, larger or smaller.

Irrational numbers are open-ended and imprecise, infinite and intangible, so that in physical form they can be only approximated. They cannot be written as fractions of whole integers nor as terminating decimals, so they lack the absolute precision of rational numbers, instead wandering off with an ellipsis-like notation at the end, like an unfinished tale, as in 6.09867... or 72.55320... Important irrationals that recur in many applied formulae include square roots, such as the square root of two, 1.414215..., or Pi, 3.141592...

We humans have devised numbers (and the written numerals that represent them) as a shorthand that allows us to talk about not only single things but multiples of things in a single group, as one entity. This codified system spares us having to count every item we discuss or even needing to see all the items of a group in the same place at the same time, bundled together. But nature has no symbols for counting: only the manifest reality of what exists. And nature, always moving, achieves addition by insinuating the new amongst the others, by adding to the line, the cluster, or the pile, adding to the growing edge or squeezing in between and reshaping the whole. In living systems, such addition is growth.

Among irrationals, vital for life and growth, nature has favorites. Just a handful of irrational numbers hold the key to the endless forms of living cells, plants, and creatures in the world, not to mention the shapes and patterns of astral constellations and the cosmos. Examples abound of these natural patterns formed by the growth of living things, and favorite examples include the seed head of a sunflower, the shape of a pinecone, or the *Nautilus* seashell. In well-formed specimens, each displays curved lines of successive elements arranged with beautiful precision in regard to number, shapes, and angles. If you look again at your hands, which we studied in the Introduction for their symmetries to one another, you can see how your own body participates in nature's predilection for the spirals and curves that are products of irrational numbers, golden section proportions, and biotensegrity architecture. Your fingerprints whorl; your finger segments, increasing in length from knuckle to knuckle, can curve around a pen; and your clenched fist rolls your hand into a spiral.

On closer examination, when counting the seeds on a seed head, scales on a pineapple, or bracts on a conifer cone, you will find the same numbers over and over. Comparing the shapes of their spirals and seeking symmetry will reveal that the angles are self-similar or sometimes exactly the same, even between very different and quite unrelated species. As the eye is drawn to these shapes, so our senses find the curves and spirals at least pleasing, if not delightful

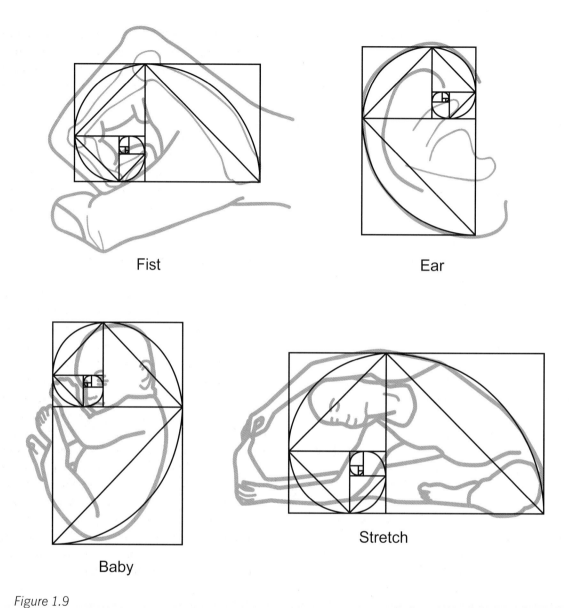

Fist

Ear

Baby

Stretch

Figure 1.9

Examples of spiral forms in the human body.

and beautiful, so they have been a perennial source of fascination and attention since human antiquity. Early cave drawings include these same spiral shapes, and ancient civilizations adopted them as symbols. Modern cultures have investigated all aspects of the art, mathematics, and science behind nature's most favored curves and spirals.

Complementary opposites

The swirling and branching patterns of living systems lie towards the richer, more complex end of a continuum of patterns based on numbers, one which starts as simply as the pattern of odd and even. These numbers alternate when we count, odd and even, odd and even, just as when we walk we alternate between right and left, right and left. Successful progression – the ability to move forward – depends on the balanced union of complementary opposites. One side of the body is not greater than another but balances in equal partnership. Each contribution matches and depends on the other's, and both are critical to successfully accomplishing the task. Balance is a foundational principle of symmetry-seeking, and we compare items for possible symmetries in pairs.

Our actions often alternate within a pair when we are making something, whether in braiding hair, stringing onions, or doing arts and crafts. We move from one side then the other, as in the warp and weft of basket making, the over-and-under of knitting needles, or the warp and woof of weaving textiles. Our actions allow us to transform flimsy fibers into stronger bundles, by twisting and winding them together, binding them into larger strands of shapes and forms. From this binding, alternating process, we make fabric from threads, garments from cotton, or roofs from straw – constructs much better suited than the original, unbound materials to meet specific functional needs. Details are different, but principles are the same, and simple repetitive actions can produce complex and variable shapes and forms.

Numbers describe and illuminate the patterns in such movements, and one of the most central uses of numbers to understand patterns is to consider the values in series. The idea of even and odd numbers, for instance, is based on the simplest series of all: the counting numbers of 1, 2, 3, 4, 5, 6.... Understood mathematically, as we look at the relationship between the numbers in this series, this simple sequence of consecutive whole numbers is based on the defined algorithm that each successive value is the sum of the previous number plus one. With the sequence established, we can deduce other rules and patterns, such as noting that our odd numbers follow even and even always follow odd, in steady alternation. And after years of familiarity with this sequence of even and

odd, we hold an intrinsic understanding of other relationships between the numbers, so familiar that we take them for granted, but which have significant consequences in application, such as that every second number in the series can be divided by 2 and still yield a whole number, but the others – the odds – cannot. Even this simple awareness of even and odd yields still further rules, such as predicting the sum of any two numbers based on whether they are odd or even: that odd plus even numbers always give odd sums, but that like numbers – odd plus odd or even plus even – always add up to evens.

These patterns based on the rhythm of odd- and even-ness have multiple implications for shapes and forms, symmetry and asymmetry, and this series is arguably the simplest possible algorithm. But nature also relies on other, more interesting sequences to define shapes and other patterns in living systems, and fractals, based on irrational relationships between numbers in a series, are the favorites. And among fractals, the arithmetic series which nature favors above all others is the one that defines the lengths of those knuckle, hand, and arm sections. This well-known series, 0, 1, 1, 2, 3, 5, 8, 13, 21, 34, 55..., is called the Fibonacci sequence, after the mathematician who first described it. Each number is the sum of the previous two, and the pairs in the sequence not only are additive but also become multiplicative: multiply any number in the Fibonacci sequence by a value called the 'golden section' or 'golden ratio', and the approximate value is the following number in the series.

Golden section

The golden ratio is 1.6180339... The decimal continues infinitely, since it is an irrational number, so we will shorten it to 1.618. It results from a formula – half the sum of one, plus the square root of five – and it can be represented by the Greek letter *phi*, or Φ. This number is also called the golden section, because it describes a unique relationship between sections of a line – one which turns out to be based on the numbers in a Fibonacci sequence.

There are innumerable ways to divide a line into two lengths and show the ratio between them, but the golden section divides a line into three elements such that each section is always in constant proportion to the others. The golden section describes that uneven cut in a line that puts the whole line, the longer part, and the shorter part in continuous geometric proportion to one another. This golden cut is the only way to divide a straight line that generates a continuous geometric proportion instead of a simple ratio.

If we give the longer part of the cut line a value of 1, we find the length of the shorter is the irrational number 0.618... The length of the whole line, then, adds the two values together, 1 plus 0.618..., yielding our golden irrational number of 1.618... By definition, subtraction reveals a difference between the whole length and the shorter to have a value of unity or 1, but, amazingly, the whole length (1.618...) multiplied by the shorter length (0.618...) also gives an exact value of 1. And if that does not strike you as strange enough, consider another curious relationship between these numbers: if you square the value of the whole line, thus multiplying this irrational number of 1.618... by itself, your result will be 2.618... – the same as if you had simply added 1 to the length.

Each Fibonacci number in the series is also an approximation of the geometric mean of its two neighboring numbers. As numbers increase in the series, the ratio of successive values alternates, oscillating to be just a little more, then a little less, but always converging ever closer to the golden mean (yet another name for this same lustrous number).

It turns out that there are many different ways to define the cut that divides the line, which will ultimately create the golden section. While the Fibonacci series is by far the most commonly occurring basis for the proportions found in nature, there are other numerical series that also appear regularly as the mathematics of living systems. A remarkable example is the Lucas series, of 2, 1, 3, 4, 7, 11, 18, 29, 47..., named after the French mathematician Edouard Lucas who described it. All such additive series (known as generalized Fibonacci sequences) behave the same way – start with any two numbers, and over time they will converge towards the golden ratio – but the Fibonacci and Lucas series get there quicker than any others. And the simple rules that define the properties of these series lead to complex consequences, their real magic becoming evident when we move from line to plane to three dimensions. The relationships between the numbers play out in the geometry of shapes and angles, the spirals and fractals we find in nature, or even in music: the Lucas series describes the mathematics underlying many well-known music compositions, such as Beethoven's string quartet Op.18, No.4 in C minor, Mozart's string quartet No.16 in E-flat major, Bach's Fugues No.11 and No.16, and Chopin's Nocturne Op.48, No.1 in C minor. All reflect the mathematical wonders of these familiar proportions, but in sounds.[5,6]

The figures, forms, and images based on golden-section measurements reveal the power of the proportion between extreme and mean: that when any line is divided in two, such that the lesser part is to the greater as the greater is to the sum, the unbroken length and the two parts of the whole provide a triad that – in living things, with their propensity to motion – seems unable to resist spontaneous assembly. The extreme and the mean are the two parts that together make the whole – complementary pieces from opposite ends of the same line.

The music of Beethoven

The many shapes built on the golden ratio show a harmony and orderly balance in their construction that echoes the harmonies of music. In fact, the vibrations in our familiar musical scales are rich with mathematical proportions, and agreeable harmonies use melodic arrangements that rely on familiar proportions, resonant octaves that reverberate throughout the whole like the architecture of biotensegrity. By now, you might not be surprised to hear that music's main building blocks, too, favor consecutive numbers from the Fibonacci sequence 2, 3, 5, and 8. On the piano keyboard, the golden proportion presents within each octave: 8 white keys, with their corresponding hammers on wires, and 5 black keys, grouped together in groups of 2 and 3.

Stringed instruments also reveal scaled patterns; the violin, cello, and double bass have a shared shape and proportion but very different sizes, which then shift their range of notes and their timbre, not unlike small, medium, and large human beings. And an orchestral symphony brings together the full range and diversity of these strings, plus brass, woodwind, and percussion, to unite them around a single score. The musicians integrate themselves in the shared task of music-making, like the autonomous cells that together build a body, responding to one another in real time. (As we'll see in later chapters, our score is the genetic code stored in our DNA.) In a push-pull of regulation and balance with one another, members of the orchestra play in a wide range of scales and fixed harmonic proportions, in a kind of musical tensegrity.

For musical harmonies to fit together in an orderly way, they need to sound at the same time. Not for nothing is a section of a symphony called a 'movement', because there is no such thing as static music, a snapshot of sound apart from time. Where the harmonious shapes defined by the golden ratio are visual, appearing within space, the harmonious sounds created by the ratios in music manifest in the dimension of time, through which music moves and flows. The alive quality of music's pulsing tempo and rhythm mimic our own most vital signs of life and health, our heartbeat and pulse.

This time-defined quality of music aligns with some of the key principles of symmetry-seeking that we outlined in the Introduction – namely, the principle of developmental simultaneity and the principle of taking the long view, looking upstream and downstream for the roots and consequences of systems. Like music, our bodies have a pulse, produce reactions, and range from quiet to loud; our organs even chime in rhythmic patterns and at moments of simultaneity, like chords. In our bodies, these rhythms are biorhythms, governed by genes, but responding to conditions and fluctuating

at times especially in response to our hormones. We move and function in patterns over time, in cycles, to the beat not only of our heart but of complementary opposites, swinging like a pendulum between in-breath and out-breath, contraction and relaxation, waking and sleeping. Like the ratios across the Fibonacci sequence, we oscillate back and forth from a little too much to a little too little, but our body's inner regulatory mechanisms, as we'll note throughout the book, are always nudging us to approach as closely as possible to our own golden mean – our sweet spot of balanced function. With a plan set by the score of our DNA, our brain stem and neurochemistry monitor and adjust like our own conductor's baton, keeping the symphony of our body in tempo and on track.

Living is about action and flow, movement and beat, and our pulse and heartbeat are only the most obvious examples of the ways our bodies keep time. Living tissues exhibit all kinds of intrinsic temporal rhythms. The blinking of an eye, the pattern of our breathing, or the rumbling of the gut are local pulses of tonal vibrations that transmit forces through the push-and-pull tension of the mechanical linkages throughout our bodies. We stub our toe at one end and grimace at the other, hear a noise and stop our breath, or feel a warm breeze on our face and relax in both body and mind – often, in each case, without any conscious thought. Whether tiny subcellular and cellular impulses or longer frequencies in tissues and the whole body, living vibrations are reminiscent of the orchestral percussionist's wide-ranging assortment of drums, cymbals, and triangle.

Our biorhythms work to synchronize not only within us but with the forces around us, which also move in cycles through time – the sunshine or darkness of day and night; the broader patterns of daylight and dark over the seasons of the year, with their longer and shorter days; or the musical cues of birds and squirrels, then of silence. Like the cells within us held together and responding to one another in literal biotensegrity architecture, our bodies – and our minds – meet with, hold to, and work with other organisms and systems in the world around us, many of them also defined by the same golden ratios and tensegrity structure that define us. And we respond to one another in a push and pull as well, each of us affecting the other, every part affecting every other part.

The macro and micro of cosmic order seem not merely coincidental or loosely linked but strangely locked in step, and biotensegrity architecture provides all living systems with push-pull linkages and leverage to respond nimbly to local conditions, including light and darkness. Not only do animals synchronize with the cycle of the day, but even some plants have harnessed tensegrity for a spry response to the sun, so that their flowers will open up with daylight and close at night. Others, such as sunflowers, will turn to face the sun and follow its course across the summer sky, their internal workings, like ours, linked to the day's tempo. Our human bodies and internal biorhythms also resonate with the pulse of the earth about us; daily rounds of circadian hormones are in harmony with the round of the sun, just as ocean tides run in phase with the waxing and waning cycles of the moon. Science continues to expand our understanding of how we are linked to the beat of our environment: the 2017 Nobel Prize in Physiology or Medicine was awarded for research on circadian rhythms, demonstrating that we have inner clocks that adapt our physiology with 'exquisite precision' to the varying phases of the day, managing 'critical functions such as behavior, hormone levels, sleep, body temperature, and metabolism'.[7] Set by genes to synchronize with the sun, the ticking of our inner clocks is the music that lays down the beat for our day-to-day function.

The structure of biotensegrity architecture, light on its feet, is the mechanism that allows organisms to respond so briskly to both inner and outer forces. It serves not only to hold the construction together but also to connect all the elements, transmitting vibrations from any one part to the rest, to be sensed, moderated, and coupled with others within the whole. This method of construction and architectural design allows the organism to adjust and tune its entire system.

This mechanical infrastructure can distribute forces to all elements throughout the whole organism, linking DNA, nuclear membranes, cytoskeletal microtubules and microfilaments, and membrane channels of the cell to the cell's neighbors and all of their neighbors in turn, throughout the living field. Biotensegrity architecture and its tensioned member proportions seem to provide the mechanical coupling system that could at the same time unite, adjust, and integrate the oscillator elements integrating the networks, bringing not only the one into tune with many but also the resonating harmonies and biorhythms of the many in accord with the whole.

Self-assembly

Much is known about nature's favorite patterns, the most persistent of which we've considered in this chapter: tensegrity, the golden section and Fibonacci, and pairs of complementary opposites. So harmonious are these shapes to humans that we replicate these living forms in countless structures, from music and art to the latest nanotechnologies. As we seek to understand the nature of human life and health, much still remains unknown. As not only science but also art, both health research and medicine are far from formulating concise laws for understanding all the aspects of human development and function, but we do know the broad principles, which we share with many other organisms and systems.

In nature, even inanimate objects organize themselves, when acted upon by an outside force. Iron filings dusted on a flat surface near a magnet rearrange themselves to align with the nearby magnetic poles, because the invisible magnetic force field will polarize each tiny ferrous fleck and order the whole, without thought, design, or direction. Dry sand sprinkled randomly on a drumhead will settle in tidy circles with no need for magnets, answering only a musician's percussion, the thoughtless vibration sufficient to assemble the sand in concentric rings, like a bull's-eye. Certain chemicals, too, when mixed in the right proportions in a petri dish, will spontaneously interact to generate unprompted wavelike patterns of color that spread, self-assembling into moving shapes – crescents, circles, and spirals reminiscent of natural growth patterns. The best-known instance results from the chemistry of the so-called B-Z reaction. All these examples – of filings and sand and inert chemicals – show how even lifeless systems can self-assemble into predictable patterns. In fact, mathematical modeling predicted patterning of this kind of behavior and years before experiments demonstrated it.

How much more easily, then, can living organisms grow themselves into predictable patterns. With

Figure 1.10

The B-Z chemical reaction makes wave-like shapes that mimic organic curves and natural symmetries. Belousov and Zhabotinsky were Russian chemists who first discovered this kind of oscillating chemical reaction and published their work in 1972. Most chemicals react to form a stable product, but potassium bromate and cerium sulphate mixed with dilute sulfuric and malonic acid react to form products that in turn generate the original reactants which keep on cycling, maintaining the reaction, like a pendulum keeps on swinging.

no need to wait for the outside force of a magnet, a drummer, or chemist's beaker, cells reproduce themselves into these structures now familiar to us, of pairs and golden sections and spirals and tensegrity.

A famous polymath was interested in finding out how living cells accomplish this feat. A multidisciplinary code-breaker, Alan Turing is popularly remembered as an early pioneer of the computer. Powered by electricity and algorithms, computers have made it possible to process not only infinitely more information than any human can but also more complex numbers than a person can calculate unassisted (including, for instance, irrational numbers such as the golden ratio). Turing famously used the computing power he developed to help the Allied Forces defeat the Germans in World War Two, by breaking their 'Enigma' code. Not coincidentally, he was also drawn to the challenge of deciphering nature's secret codes, of the kind most interesting to us here: he was a pioneer in developmental biology.

Turing's was a symmetry-seeking question: he wanted to know how cells could be the same, the same, the same, and then different. Specifically, he was looking for the chemical mechanism that would explain how once-identical cells in the embryo could become chemically different, after they divided and subdivided. What led to their changes in shape and size, and in turn caused cellular differentiation, assigning one fate to some cells, but quite a different fate to another: skin or bone or blood cells?[28]

When he offered his multidisciplinary explanation, Turing proposed that pairs of complementary opposite chemicals, one promoting and the other inhibiting a particular chemical reaction, would diffuse across a field of similar cells such as the early human embryo and would create distinct patterns of cellular differentiation. Insisting that 'certain well-known physical laws are sufficient to account for many of the facts', he applied the known patterns to cells, publishing calculations to suggest that six distinct patterns of cellular differentiation might be generated through simple chemical interactions of reaction and diffusion, in oscillations that cause a little more and a little less of each chemical in neighboring cells.

When technology advanced sufficiently to allow researchers to test Turing's findings, they were able to validate his predictions. Scientists recently managed to confirm all six of Turing's original patterns, plus one additional pattern not predicted by his original computer model.[9]

Notably, the patterns considered in Turing's research and its offshoots involve the same branching patterns or whorls that interest us in the human body and that appear in Bejan's work on flow – tree branching patterns and whorls – products of the golden mean.

Symmetry tells us that answers to questions about living systems lie upstream, in their development, which means answers about the human body and thus human health lie in our embryonic development. Contemporary scientists have spectacular tools and resources, far beyond those imaginable 70 years ago, to build on the work of Turing and others who pioneered modern embryology, so we now have a great deal of detailed knowledge about the development of the human embryo, fetus, and child. Still, important questions remain to be answered, about what most affects our form, function, and developmental trajectory. And while the ever-growing body of facts can aid our understanding, still more important is better synthesis of what is already known, a putting together of our knowledge across a broader range of fields. For researchers, it would make sense to consider how we might articulate better the rules of human growth and interpret the lessons of individual variations.

For health care providers and all of us who are seeking to understand how form shapes our health and wellbeing, the lesson of biotensegrity and cell structure is the interconnectedness of every part of our bodies – a theme that symmetry reasserts again and again. In the next chapter we shall explore some of the patterns of disordered human growth, along with their implications for form and function.

References

1. Thompson DW (1917). On Growth and Form. Cambridge: Cambridge University Press.
2. Bejan A (2016). The Physics of Life: The Evolution of Everything. St. Martin's Press; p.2.
3. Scarr G (2014). Biotensegrity: The Structural Basis of Life. Edinburgh: Handspring Publishing.
4. Doczi G (2005). The Power of Limits: Proportional Harmonies in Nature, Art and Architecture. Boston: Shambala Publications Inc; p.5.
5. Madden CB (1999). Fractals in Music: Introductory Mathematics for Musical Analysis. Salt Lake City: High Art Press.
6. Madden CB (2005). Fib and Phi in Music: The Golden Proportion in Musical Form. Salt Lake City: High Art Press.
7. Nobelprize.org (2017). The 2017 Nobel Prize in Physiology or Medicine - Press Release. Online: www.nobelprize.org/nobel_prizes/medicine/laureates/2017/press.html.
8. Turing AM (1952). The chemical basis of morphogenesis. *Philosophical Transactions of the Royal Society of London. Series B, Biological Sciences*, 237(641): 37–72. Retrieved from http://www.jstor.org/stable/92463.
9. Tompkins N, Li N, Girabawe C, et al. (2014). Testing Turing's theory of morphogenesis in chemical cells. *Proceedings of the National Academy of Sciences*, 111(12): 4397–4402.

Heads over tails: looking for patterns

In Scotland, legend has it that a highlander fell in love with a mermaid, and from their union was born the first MacLaren. The clan crest, coat of arms, and colors all bear the image of this familiar mythical creature, with its upper body of a woman, joined to the lower body of a fish. Growing up in Scotland in the MacLaren clan myself, I was intrigued by the story but never thought to tease out its mysteries. As a doctor, I have since learned a great deal about mermaids, including what they can teach us humans about our own bodies.

Fascination with the fantasy of partly human aquatic creatures reaches back to ancient times, and the best-known mermaid myth was published in Denmark in the mid-19th century: Hans Christian Anderson's 'The Little Mermaid', the fairy tale of a mermaid who wants to become human. Representations of mermaids have since appeared throughout literature and art, whether from composer Richard Wagner or playwright Oscar Wilde, painter John William Waterhouse or novelist Sue Monk Kidd. In recent decades, mermaid lore continues to thrive in popular culture, in stories that often cue from Anderson's fairy tale, such as the 1984 movie *Splash* or the Disney animation *The Little Mermaid*.

Curiously, the mythical mermaid is usually a sexual symbol, dangerously attractive to male humans – in contrast to real-life mermaids, which, unbeknownst to many people, do exist.

Figure 2.1

The coat of arms of clan MacLaren.

A genuine mermaid

One December about 25 years ago, a teenage mother from the Appalachian Mountains brought her newborn to our children's hospital in Atlanta. We had never seen an infant like this unfortunate baby, who had been born with the rarest of all congenital anomalies: her legs were fused together, and she had no external openings for bowel or bladder or vagina. The child was a mermaid.

This baby's condition represents the most severe form of neural tube abnormality possible, one that typically provokes miscarriage. An infant with mermaid syndrome that does survive to birth is not likely to live more than a few days. In this condition, called sirenomelia (after the sirens of Greek myth, whose history is interwoven with mermaids), the nervous system has no tail end, so there are no nerves to the legs. Lacking nerves, the legs fail to form muscles, so there are no forces to move the lower limbs apart. Instead of becoming two legs, the lower body remains flaccid and lifeless, wrapped in a single sheath of skin. Thus the infant with mermaid syndrome appears to have a tail, like the mythical mermaid. In reality, however, the usual tail end of the body – including the tail end of the nervous system – is simply missing, leaving the inert limbs fused, in the appearance of a tail. The triangular bone in the pelvis (the sacrum) and what we call our tailbone (the coccyx) are the bony structures at the base of the adult human spine – our tail end – and, like the rectum and anus, vagina, urethra, and bladder, these bones never form in the mermaid infant.

Sirenomelia is so rare – 1 in 100,000 live births[1-3] – that none of us modern doctors had ever seen an infant with mermaid syndrome until that December day, even though we were in a major teaching hospital. But when I saw this baby, I immediately remembered my ancient MacLaren clan history. With four children of my own and, so far, eleven grandchildren, I know how readily we attribute some feature of a newborn to one relative or another – a nose like Uncle Hamish or ears like Auntie Joan. So the infant mermaid made me wonder whom she looked like. Faced with such a baby, our ancestors, too, no doubt looked for some kind of connection and, not finding one, might have created a tale to explain the mystery of such a child.

If the baby had been born into a primitive community in the ancient land of Alba, as Scotland was once known, what better way to help its family explain the birth of such a baby than a mythical link to some remote ancestor – a wondrous progenitor endowed with mysterious powers, better still. Many of those same Highland Scots emigrated and live in the Appalachian Mountains of north Georgia, Tennessee, and North Carolina, and we now understand the role of genetics and environmental factors in these abnormalities better than we did twenty-five years ago. So the possible connection between that unfortunate adolescent mother, her newborn baby, and the mermaids of the MacLaren crest came more clearly into focus. Perhaps, somewhere in my family's history, there was a mermaid.

Unaware of the romantic fairy tales surrounding semi-human sea creatures, the baby in the children's hospital had, in addition to her fused lower limbs, severe cardiac and other abnormalities that were not amenable to surgical correction. She died peacefully when she was three weeks old.

What rare tells us of ordinary

The sex-symbol status of mermaids in mythology contrasts starkly with the reality of the real human mermaid, who has no sexual structures and no possibility of a sexual life.[4] In mythology the one-eyed Cyclops is endowed with special powers directly related to its anomaly, and it also appears in science fiction. But in the very rare event that a human is born with only one central eye, the infant is always blind.[5,6]

On the other hand, science fiction sometimes portrays stark human weaknesses as counterbalances for new strengths, such as humans in the distant future with evermore powerful brains but weaker human bodies – often without a capacity for moving about without the aid of an electronic transporter, such as the Daleks on *Doctor Who*. Curiously, this trade-off between strength in the upper body and corresponding weakness in the lower body is actually truer to evolutionary reality than the myths of superpowers, and it's relevant to what we can learn from the mermaid, as this book will explore.

The mermaids and their kin of legend lured men to their deaths, their sensuous beauty an illusion which masked their actual sexlessness. But though the real-life mermaid of sirenomelia is not only sexless but sadly harmless and inert – unable to walk, indeed like the mermaids of legend – of what other danger might she warn us? What does the mermaid tell us about ourselves, about human form?

In all human bodies, nerves grow out from the spinal cord to each limb and organ. The human mermaid has enough nerves to make some kind of legs, albeit fused together, but not enough to make external genitalia or a sacrum or pelvic floor. She is unfinished. The tail end of her nervous system is missing.

That the legs are slightly more finished than the genitals reflects the human body's developmental priority: the nerves to the legs and feet grow out from the spinal cord segment closer to the brain than do the nerves to the pelvic floor and anus. Legs seem to be lower on the body than genitals and our tail end, when we look at a human standing with feet on the ground, but, viewed from the perspective of the spinal cord, the nerves to the lower extremities – the legs and feet – are actually higher up than the nerves to the genitals. (It's easier to see where they connect up and down the spine if you look at it from the back of the human body, and sitting, as we'll see in Chapter 3.) And when nature fails to make a perfectly formed spinal cord, the parts of the body furthest away from the brain will be more imperfect than any others. So structures in the pelvis, at the tail end furthest from the brain, are the most likely to suffer from incomplete development.

Nerves, along with special builder cells called neural crest cells, also play a key role in tissue growth and development. They are like the foremen on a building site who direct the project and make sure that the form of the structure is an accurate reflection of the body plan, which is held in the genetic code. Nerves are thus critically important in the embryo, to direct growth and form. The brain is the control center and the spinal cord is an extension of the central nervous system that extends to

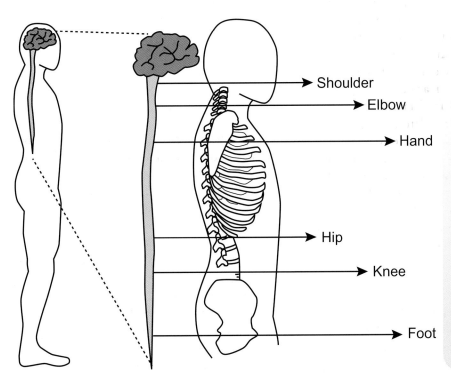

Shoulder
Elbow
Hand
Hip
Knee
Foot

Figure 2.2

How the nervous system 'views' the body. Functional organization of the spinal cord as it descends from the brain: the first nerves control the muscles that move the proximal joints of the limb – shoulder muscles first, followed by those that move the elbow, then wrist – and the last nerves move the most distal joints of hands and feet, fingers and toes.

reach the body parts. At the end of the spinal cord, nerves give way to fibrous cord, like a road whose paved surface ends and becomes only a dirt track. Where that end happens – and whether the paved road reaches far enough to support all the segments of the human body – turns out to be critical for human health, especially in the pelvis.

What value in a tail?

Most mammals have a tail. Other than humans and great apes, almost all of them have this extension of the body beyond the anus. And because the nerves of the spinal cord extend into all the body parts including the tail, the mammals with tails have more nerve segments in their spinal cords than humans have, segments which connect to and control the muscle groups that move the tail. While in some larger mammals, such as monkeys and cattle, the tail has clear uses – in signaling, balance, locomotion, and brushing flies away – more often it seems to have no obvious primary function, especially in smaller animals. Thus, a tail might be somewhat useful for the cat but only burdensome for the mouse.

It would be strange for a body part that is present in nearly all mammals to have such limited usefulness, so, looking through the lens of symmetry and knowing nature's relentless economy, this lack of obvious function prompts the scientific mind to ask what more fundamental purpose a tail might possess. Logic would suggest a greater purpose for the tail than swatting flies. As nerve tissue becomes thinner and thinner when it approaches the extreme tip of the spinal cord, the nervous system is drawn out to an end in the tail. So a tail, by giving extra length to the spinal cord beyond the vital body segments such as legs, feet or pelvic organs, creates a safety zone that protects the body from the effects of incomplete spinal cord development.

Thus, in all the animals that have a tail, it offers a universal protection for the development of the nerves for the lower extremities (the legs and feet) and for the structures in the pelvis (the urethra,

vagina, colon, and anus). By extending the nervous system beyond the trunk of the body, a tail shelters the nerves of the lower body during neurological development, providing a safety zone that protects against developmental shortfall. Species lacking a tail – humans, the apes, and a few other rare mammals – are therefore peculiarly susceptible to problems in the tail end of the body.

As evidence of this problem with our lower body structures, too many human babies are born with spina bifida, a condition involving loss of spinal cord structures, to varying degrees of severity. Animals that have a tail, such as mice, rats, dogs, horses, and cattle, are never born with spina bifida, so these other mammals have not offered helpful animal models for studying the condition. While scientists can rely on animals when studying many human diseases, the typical features of spina bifida appear only when a mammal has no tail – a condition which, as we have seen, is rare. So while human diabetic mothers are at greater risk for having a child with spina bifida, diabetic mice studied through pregnancy will simply produce pups with normal-looking bodies but abnormally short and curly tails, not spina bifida.

Manx cats are probably the best-known example of a near tailless mammal. The severity of their tail-lessness covers a wide range, from 'stumpies', that have a rudimentary tail, to 'rumpies', that have no tail at all. This cat has proved useful for researchers because, like the human, it can give birth to offspring with spina bifida. Some move about like normal cats, but others have a curious bobbing gait, like rabbits, and some – called hoppers – are paraplegic and can only hop, since they are paralyzed from the waist down. Manx cats also have a rather large head and a reputation for intelligence and ease of training.[7-9]

Given the prevalence of skeletal deformities among Manx cats, especially vertebral and tail-end defects, responsible breeders do not allow them to mate with one another. They often have bowel and bladder incontinence, incomplete bladder emptying, urinary infections, and bladder stones, and kidney failure often causes their premature death.

Figure 2.3

Manx cat. They may have a short stumpy tail or no tail at all. The back legs are often spastic or clumsy.

Figure 2.4

Human embryo at 5 weeks, total length 5.5 mm.

Like humans, but unlike tailed mammals, some Manx cats are born with an anus that is sealed up at birth (imperforate) and must be opened surgically for the animal to survive.

Human tails

While a look at the outside of our human bodies shows no tail, the perfectly formed early human embryo does in fact have a large and distinct tail. The form of all mammalian embryos, across species, is very similar in the first weeks after conception. Nature has one pathway for building a mammal, from a single fertilized egg cell to a ball of cells and then into a disc-shaped embryo. The disc has a head end and a tail end, and the tail of the human embryo is most prominent in the fifth and sixth weeks after conception.

Perhaps paradoxically, cell death and tissue destruction turn out to be almost as important for building an embryo as cell growth and proliferation. Just as in a construction project, where lumber must be cut to size to reach the right proportions, so the

human tail is trimmed away, when the embryo is no more than 6 millimeters long.[10] Like Grandma making a gingerbread man, nature rolls out more tissue than she needs for making an embryo. But while Grandma uses a mechanical tool – the cookie cutter – to cut clean edges as she trims away excess dough, the embryo has only chemical signaling to define how much will be lost and what will remain. In an embryo, instead of an absolutely constant, reliable, reproducible measure that would produce cookie-cutter tail ends, there is variation: instead of a constant and defined, fixed edge, there are transition zones where living tissue gives way to dead.

Unlike the clean-edged cookie cutter, chemical signaling is imprecise, more like a tide on a beach. The spring tide of cell death advances, washing away unwanted tissue from the tail structures of the embryo, while leaving some untouched. But because

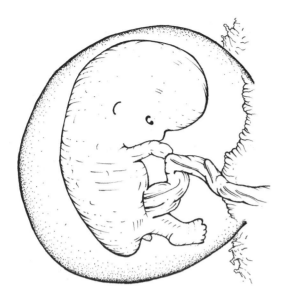

Figure 2.5

Human embryo at 8 weeks, total length 6.5 mm.

are more similar to each other than one's feet. We see the same pattern between one human body and the next: more structurally similar to one another in the upper body, more variation in the lower. As evidenced in mermaid syndrome, neural structures taper off in their successful development, with the greatest loss most distant from the brain. Even within the tailbone itself, deficits at the tail end will be more common, with problems further up the spine and spinal cord occurring more rarely. We can afford more structural divergence at our bottom end than in our brains.

In a medical school anatomy class, the students each receive a human skeleton, and together they examine how bones of the skull or the thighbones are constructed. But they never bother studying the tailbones together, because the wide variation in the tail end of the human body makes each sacrum and coccyx so different from any other, across the wide range of variation that encompasses both congenitally deficient and fully formed tailbones, that they are not consistent enough to examine simultaneously.

Age amplifies initial defects

The most severe congenital deficits include multiple abnormalities – usually, problems in more than one system in the lower body – while milder forms of deficit produce fewer abnormalities, sometimes minor enough to be invisible. Sooner or later, though, and sometimes much later in a person's life, the impact of deficits in the tail end of the body will manifest as imperfect function of the pelvic organ systems – the bowels, bladder, and sexual organs – even if the roots of these problems are invisible, because they began decades earlier, in incomplete embryonic development.

the high-tide mark of cell death is not the same for every embryo, this mechanism is unable to act with enough precision to delete only the unwanted excess while always preserving the full complement of structure every time. At times, the tide of cell death will overshoot the perfect place, eroding some needed tail-end structures and washing away some parts that ideally would have remained untouched. And the consequences of this event, seemingly trivial in the embryo, amplify later, as growth exaggerates the penalties of missing the mark, creating measurable effects on the form and function of the most caudal (furthest towards the tail end) segments of the human body.

Erosion of the lower body segments that results from washed-away tail structures can occur in either sex, to varying degrees of loss, across a spectrum of deficits. But as we look for a pattern across these differences, it appears clearly: top to bottom. The upper body is more symmetrical, becoming less so further down – so that, for instance, one's hands

In animal embryos, incomplete development of nerves means a well-developed animal with an incomplete tail; but in our human form, unprotected by a tail structure, incomplete development of the caudal (tail-end) nervous system has a multilayered impact on the body and causes multisystem problems. More severe losses affect not only bowel

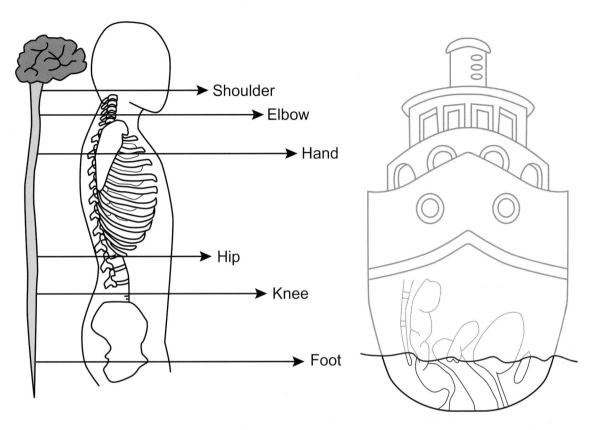

Figure 2.6

Symmetry of the human nervous system, and a ship in the water. The lowest parts are hidden down below out of sight, but critically important for the functioning of the whole.

and bladder, penis and vagina, but also the form and function of the toes, feet, and lower limbs.[11]

Nerve cell populations are laid down in sheets, stacks, and layers in the brain and spinal cord, each reaching out like tiny tube-like telephone cables to make contact with other nerves, receptors, or muscle fibers throughout the body, clumped together in groups or strung out in chains. The number of nerve cells distributed in the developing fetus will not increase much after birth, and, after the second decade of life, a slow but measurable loss of nerve cells begins. This process of attrition is one factor that contributes to the slow erosion of structure and a corresponding loss of potential for full function later in life.

In ordinary aging, we are accustomed to the visible patterns of loss to outer structures: loss of hair, thinning skin, and diminished strength and muscle mass. But parallel losses within the body have a profound impact on overall health, especially because they tend to involve further erosion of nerves in the tail end of the nervous system, a washing away that began in the embryo but whose effects might not become evident until aging continues the process.

Medical students are taught to distinguish between diseases that arise before birth and those that come later; doctors call them either 'congenital' or 'acquired', understanding a congenital condition to mean that something was not made the way it should have been. A baby with 'a hole in the heart' – an abnormal opening between the right and left sides of the heart – has a 'congenital' heart disease. 'Acquired' on the other hand, means that everything was well until the problem occurred – perhaps arthritis of the joints or kidney disease. In reality, though, most diseases are both congenital and acquired. The world is not flat – we are not all the same – and variation between individuals makes some of us less vulnerable to certain conditions and others more susceptible to disease or dysfunction. That symptoms of a given condition may not appear until later in life doesn't mean that the underlying vulnerability hasn't been with us since before birth.

The gift that the perspective of symmetry brings to our health is a way to discern the shape of our own particular part of the un-flat landscape, literally the shape of our own bodies, both externally and internally. With an eye for pattern and knowledge of where human evolution has brought us – and where it is carrying us – individuals making personal health decisions can better understand what is before their eyes, as can health professionals making diagnoses and guiding treatments. With this perspective, we can recognize and minimize the impact of our own congenital vulnerabilities, thus better managing or even avoiding some acquired and seemingly unpredictable problems. Recognizing symmetry, we learn better what to expect.

Evolution is change over time, and it is easy to see the positive advances at the head end, with our modern human brains having grown considerably larger than early humans, *Homo erectus*. It is also easy to understand a larger brain as better than a smaller one, with more brain cells, more cognitive ability, and greater ease adapting to and succeeding in the world. But in our enthusiasm for the wonders of adaptation, we sometimes forget that change is not always beneficial. Variation brings random changes, and only interactions with an organism's surroundings determine whether the adaptations offer advantage or disadvantage for survival. Over eons, the variations and then prevailing conditions have together defined which changes flourish: which are adaptive, therefore

Impact of aging – shift from left to right →

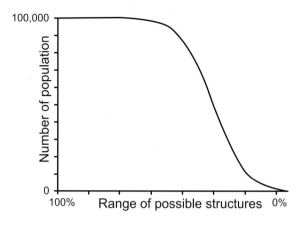

Figure 2.7

Human variation in distal spinal cord innervation. This graph reflects the range of possible nerve density in the distal segments of the human spinal cord. Many of us are blessed with complete structure and full versatile function because all the nerve cells are fully formed and functional, as illustrated on the left side of the graph – top of the hill. Others are born with fewer nerves in the tail end of the spinal cord represented by the downward slope, and some with severe congenital deficits have hardly any nerves in this area at all. No nerves are represented on the extreme right side of the graph.

Over years of aging or with losses from injury, inflammation or disease, further loss of nerve cells may occur but the impact of those losses might be very different from one patient to another. Those with a full complement of nerve structures can afford to lose some without significant impact on function, because they start with such a wealth of structure. Others in the mid-range may have little or nothing in the way of problems in early life, only to have more in later life: a shift will occur from left to right with further loss. Some drift to the right is inevitable as we age and go from having everything we started with to having less. The magnitude of loss can be similar for any two patients, but the impact of losses might be felt very differently according to the initial conditions.

persisting across generations, and which fail. If, from generation to generation, some changes in form overshoot the sweet spot of optimal advantage, then instead of adaptation they create real disadvantage – obstacles – and for humans, as we have seen, the loss of tail structures is one of those maladaptive developments, even counterbalanced as it is, overall, by greater development of the brain.

Evident erosion of tail-end structures, ranging from mermaid syndrome to spina bifida and imperforate anus, is an obvious example of maladaptive human development. But the best example of the problem that evolutionary change has created for our heads and tails lies in the everyday event of childbirth.

Obstetric dilemma

Head first into the world as a baby is born, the crowning cranium leads the way in ninety-five percent of deliveries. Shoulders and body follow, with legs and feet dragging behind. Even from the ordinary beginnings of life, then, heads lead and tails follow. Choosing heads or tails, nature favors heads at every opportunity, every coin toss, always tending to apply more resources towards the head end at the expense of continuing cuts and economies at the tail.

Blessed with the largest brain and therefore the biggest fetal head of any creature, in proportion to our body size, we must be pushed and pulled through the birth canal to enter the world. Unlike other mammals' childbirth, ours carries special risks for both mother and child, and vaginal delivery is the most dangerous journey of our lives. While for other animals birthing is usually mundane, for us humans it is more often a suspenseful drama, in which the disproportion of an exceptional head is molded by the resistance of an unyielding cervix and pelvic floor, the trapped fetus arrested too often in mid-passage, unable to escape without surgical incision or the mother's tearing open, or

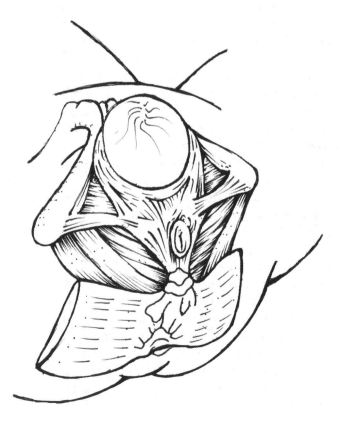

Figure 2.8

Vaginal delivery with the baby's head crowning and stretching the perineum and muscles of the pelvic floor. Sometimes it is necessary to cut the thin rim of vaginal tissue to help release the baby's head and allow delivery to proceed. This is called making an episiotomy.

the use of obstetric forceps. Tangled in the cord or stuck in place, our babies can even lose blood flow to the brain and suffer the permanent injury of cerebral palsy. No longer just natural, childbirth often becomes an intensive medical experience, supported by teams of nurses and doctors with drugs and procedures, monitors and instruments, all ready if needed to surgically extract a distressed fetus by caesarean section, to protect the lives of the mother and newborn. For many women, childbirth proceeds easily, even joyfully, and does little discernible damage to the pelvic floor. But others suffer profound consequences, some of which may be evident immediately, others lasting a lifetime but perhaps not evident until later, such as pelvic support anatomy disturbed, changing bladder or bowel control for the worse.

Only humans have to struggle with the obstetric dilemma, trying so hard to give birth or to be born, because it is in childbirth that evolution's triumphant head faces opposition from a sometimes obstinate and unyielding pelvic floor at the tail end. But the lens of symmetry, starting with the simple inspection of their own bodily structures, can help women and their doctors better understand their individual strengths and risks, to guide their decisions about pregnancy and childbirth.

Consider anatomical structure as simple and obvious as a person's feet and toes, which offer a telling indication of a woman's likelihood of having well-developed structures for vaginal delivery. If her feet are well-formed and versatile, her pelvic floor has better odds of being able to cope with childbirth.

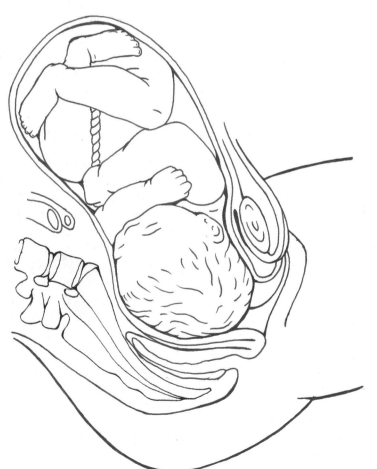

Figure 2.9

Vaginal delivery. The birthing baby typically presents with the back of the head pointing forwards. Baby's head must squeeze through the mother's pelvic floor in a manner similar to pulling on a T-shirt.

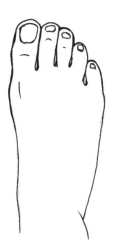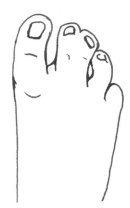

Figure 2.10

Human feet are very varied in shape and form. A fully formed versatile foot (left) is neither high arched nor flat footed, toes are not partially nor fully fused together and toes do not deviate much from the line of the others. Feet are often included in art and sculpture providing an excellent opportunity to observe patterns of variation.

Flat-footedness, feet of different sizes, or toes that will not spread apart indicate a less-versatile pelvic floor, one less able to function optimally during vaginal delivery, because, given their place in the hierarchy along the spinal cord, tail-end structures suffer even more from developmental limitations than feet.

Her personal health history, including information about childhood experience with bowel and bladder, can also guide a woman and her doctor in making decisions before and during childbirth. And because we inherit our structures from our family members, the birthing experiences of mothers, grandmothers, and sisters may reflect an increased likelihood of certain childbirth events for a given woman. Knowing that human development follows identifiable patterns – prioritizing the head and upper body first, to be more symmetrical and balanced, and the lower body last, more likely to be asymmetrical and unfinished – we can see health problems formerly thought unrelated to one another actually lying along this spectrum of neurological deficits. In my work as a urologist, I have found this application of the idea of symmetry – of looking for these kinds of similarities and balances, between systems – profoundly useful in treating my patients and understanding their health risks.

Symmetry informs personal health

Knowing the times and the ways that we are most vulnerable to incomplete development or to neurological damage can change the way we manage our health. Not only do genetic instructions lay down our plan for individual features, across a range of discernible variations, but the initial conditions of embryonic development prove to be critical to future health, in heretofore unknown ways. If we note certain milestones during childhood and certain events at pivotal junctures in adult life, some of which reflect those embryonic developments, they too can guide our choices as we navigate the aging process. Science now offers each of us not only an untapped reservoir of potential information about our personal health trajectory but a mine full of data about how the conditions of our daily life, especially the food we eat, guide that trajectory towards its worse or better possibilities.

Symmetry-seeking offers a path for making sense of this wealth of data. It is a learnable habit of looking for the specific, big-picture similarities between one part of the body and another, between one human and another, between one species and another. Looking beyond the silos of their own specialties, health professionals can see connections across body structures and systems that can guide new approaches to diagnosis and treatment. Better still, the people most able to affect our health positively – our own individual selves, making day-to-day choices – can apply the lens of symmetry to look for better ways to prevent illness, minimize its negative impact on our lives, and find a path to overall wellness, which doctors cannot find for us.

Applying symmetry to the broadest perspective, we also discover where asymmetries and imbalances lie and find ideas about how best to respond, not only at the level of policymaking, but on trips to the grocery store. As the coming chapters will explore, an eye for symmetrical patterns show us why hard-to-avoid environmental conditions that affect us all have a particularly profound impact on the unborn, during embryonic development. With a better understanding of the problems and their causes, we can then focus critical attention not only on helping buffer negative conditions for pregnant women but on changing problematic conditions for everyone, whether in our systems for delivering health care, our medical institutions and training, our food supply, or our contemporary uses of pharmaceuticals. We can thus increase the range of healthy, even lifesaving choices for both the old and the young, faced as they are today with the threats of obesity, diabetes, and cancer.

Our mermaid holds up a rare mirror to reflect asymmetry, marking one tragic end of the spectrum of human possibilities. The young teenaged mother from the Appalachian Mountains, who was still growing her own body while sharing poor nutrition and limited resources with an unexpected embryo, had little hope of overcoming unknown developmental dangers to her fetus. Unlike the adult mother-to-be who is already fully formed, the adolescent frame is still pulling obligatory resources for growing muscle and bone at the expense of failing to feed the needs of her embryo's growth and development. As you read about medications, hormones, and food in the coming chapters, you will see the profound opportunities to apply that knowledge towards better health: the way our choices in all these arenas fit into a grand symmetrical pattern, at this moment in our human evolutionary history.

Ultimately, the great human head may be our big obstacle during childbirth, but put to good use – to its highly evolved and adaptive use as a collector of data, a source of new ideas, and a maker of meaning – the human brain is giving us the vital resources necessary to help us navigate and negotiate all the many forces at work for and against health and wellness.

References

1. Orioli IM, Amar E, Arteaga-Vazquez J, et al. (2011). Sirenomelia: an epidemiologic study in a large dataset from the International Clearinghouse of Birth Defects Surveillance and Research, and literature review. *American Journal of Medical Genetics*, 157C(4): 358-373.
2. Kshirsagar VY, Ahmed M, Colaco SM (2012). Sirenomelia apus: a rare deformity. *Journal of Clinical Neonatology*. 1(3): 146-148.
3. Das SP, Ojha N, Ganesh GS, et al. (2013). Conjoined legs: Sirenomelia or caudal regression syndrome? *Indian Journal of Orthopaedics*, 47(4): 413-416.
4. Stahl A, Tourame P (2010). [From teratology to mythology: ancient legends]. *Archives de Pediatrie*, 17(12): 1716-1724.
5. Olejek A, Bodzek P, Skutil M, et al. (2011). Cyclopia – literature review and a case report. *Ginekologia Polska*, 82(3): 221-225.
6. Cannistr C, Barbet P, Parisi P, et al. (2001). Cyclopia: a radiological and anatomical craniofacial post mortem study. *Journal of Maxillofacial Surgery*, 29(3): 150-155.
7. Deforest ME, Basrur PK (1979). Malformations and the Manx syndrome in cats. *Canadian Veterinary Journal*, 20(11): 304-314.
8. Green ST, Green FA (1987). The Manx cat: an animal model for neural tube defects. *Materia Medica Polona*, 19(4): 219-221.
9. Leipold HW, Huston K, Blauch B, et al. (1974). Congenital defects of the caudal vertebral column and spinal cord in Manx cats. *Journal of the American Veterinary Medical Association*, 164(5): 520-523.
10. Moore KL, Persaud TVN, Torchia MG (2011). The Developing Human: Clinically Oriented Embryology, 9th Edition. Philadelphia: Elsevier.
11. Stec AA (2011). Embryology and bony and pelvic floor anatomy in the bladder exstrophy-epispadias complex. *Seminars in Pediatric Surgery*, 20(2): 66-70.

Not just parts:
look at the whole body

When Leonardo Da Vinci drew Vitruvian Man, he provided a template for symmetry-seeking, because he looked for symmetries all over the body, moving the axis to multiple different heights and angles, to see what mirroring he would find. The full-on torso, face, and midline parting of the hair emphasize the vertical midline axis, but you can reveal more – learn more from symmetry – by playing with the placement of the axis, not only right and left but top to bottom. Da Vinci was able to highlight new symmetries along the horizontal midline axis. By lifting the figure's arms and spreading its legs to touch the circumference of a perfect circle, Da Vinci contains the figure such that all diameters course through the navel, intersecting at the very epicenter of the circle. The new figure invites us to see the sameness of not just ten fingers and ten toes, or arms and legs, but also of shoulders and hips, or of thoracic inlet and pelvic outlet.

When you look in the mirror, right-left symmetry is easy to see, and we sometimes notice other obvious similarities, such as the top-bottom resemblance between our hands and feet. But our bodies are filled with a host of fascinating symmetries, some of them subtle, obscure, or altogether hidden in our insides. It's time now to delve into some of these patterns, to consider how the body is made, where other symmetries might lie, and how they echo the ones we can see. Following Da Vinci's example, our symmetry-seeking will involve moving the axis to new locations and angles, to see whether previously unnoticed symmetries emerge.

Figure 3.1

Vitruvian Man in a circle. The umbilicus is the perfect center of the circle; all diameters drawn pass through the center. A transverse diameter provides an axis/horizon with upper body above and lower below. Redrawn by Paul Rodecker.

Different and same

Difference is easy to spot. To call two items different, we need find only one contrasting feature and the case is made. And having found the difference, we may stop looking for similarity. Even highly educated experts tend to recognize difference more readily than similarity, for specializing often involves intense study on a narrowing focus in a chosen field. Such hyper-focus tends to exclude the much wider, broader view that may be needed to see where similarities might be found.

Symmetry teaches us to look beyond those obvious differences for similarities, and some of them are easy to spot, too, once we train our attention towards them. But at the heart of this method that we call symmetry-seeking, it's important to keep looking beyond one or two obvious similarities, because it takes a plurality of shared traits to claim symmetry, of the kind that deeply informs understanding. Having experimented with the placement of our axis of symmetry and noted a similarity, we must stay on the trail, asking the questions raised by the symmetry-seeking principles, to consider origins, timing, and development, so that we can unearth the clues and connections we need to comprehend the whole.

As you follow the trail with us to see how we apply the principles to human organs, growth, and development, it will become clear that similarities are much more frequent than differences in the body. The similarities may be covered up, like our feet, hidden in shoes, while our hands remain visible, and we may notice only the differences – between feet that go unnoticed unless they hurt, while hands perform visible wonders of fine motor movement, holding pen, brush or scalpel. But, structurally and developmentally, the similarities between arms and legs, hands and feet, or fingers and toes turn out to be much more profound than the differences. They're similar enough that a foot can lend a digit to an injured hand: transplanted by a surgeon, the big toe can substitute for a lost thumb. While not an exact match, it's enough like a thumb to restore a functioning grip.[1]

Patterns of growth and development are the same between hands and feet because the body-plan blueprint is the same for all limbs, sending out the same instructions for the cascade of genetic signals

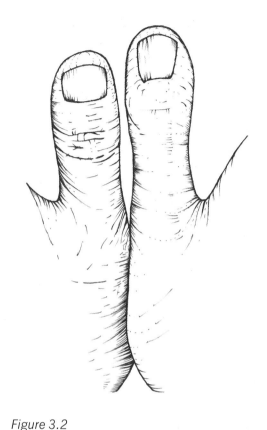

Figure 3.2

Big toe to thumb transplant. Mirror anatomy means that all of the structures of thumb and big toe – bone, arteries, veins and nerves – will fit together, like with like.

that direct a limb to grow. The processes that lay down the positions and identities of individual cells and direct their different fates are quite the same whether right or left, fore limb or hind limb. And the result of all this sameness is symmetry. The distribution of bones, joints, muscles, fascia, blood vessels, and nerves is organized in the same way throughout the body, and the similarities allow anatomy teachers to rely on those patterns when students are learning about limbs. Once you understand the layout of the upper limbs, you already know a lot about lower limbs and can predict how they grow and function.

Not just hands and feet but arms and legs are also symmetrical with one another: similar but not the same. Scientists have now studied the details of the growing embryo thoroughly, and the most remarkable discovery is that all embryos are essentially the same. There is only one path of progress from fertilized egg to embryo to fetus. Nature offers no alternative means of development: just one obligatory, universal, and well-trodden path. No longer a mystery, molecular biology has revealed the precise mechanisms of development and its sequence of key steps. Humans also share this path with other mammals, which is why animal studies can inform our conversation about the human condition. Now we know the triggers for growth, and the greatest surprise is that growth needs only one shared set of genetic instructions – one toolkit for all that is fundamentally the same across countless species.

Cell and whole-body symmetry

In humans and all animals, a single cell is simultaneously the basic building block for constructing the body and a self-contained template for the whole. As they multiply, self-organizing cells pack themselves together into tissues. Unlike oranges that passively fill a box, laid one after another in rows and stacked in layers, living cells burst out from within to create more cells. Powerful forces are at work, prompting one cell to become two, two to become four, continuing on as the cells squeeze against each other, filling up space and shaping the body. Each cell is held in an architecture of biotensegrity, holding sway over its neighbors, and each cell participates in a reciprocal dialogue of push-and-pull forces, with both local and remote influence on different regions of the body, all while nerves, hormones, and other biochemical gradients send signal messages to all of the cells at once.

As we saw in Chapter 1, biotensegrity holds each cell together, in an outer membrane that encloses this microscopic bag of powerful contents. Tiny

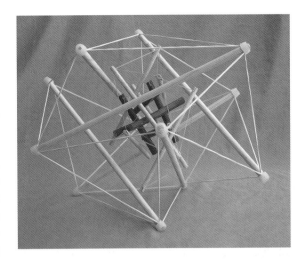

Figure 3.3

Biotensegrity: struts and strings cell model.

tubules frame the inner contents, like poles of a tent or spokes of a wheel, to keep everything orderly, attached where it needs to be, and held in place.[2]

Picture a spider's web with its catch of flies; there are such webs within the cell, holding tiny sub-cellular organs in place – a microscopic preview of the way the human body will soon position and support the kidneys and liver. In the cell, Lilliputian strands hold and tension networks of tiny tubules, extending all the way to the cell membrane. These tiny tubules and microfilament strands perform important functions both within the cell and to and from neighboring cells, transporting chemicals and push-and-pull messages.[3]

The tubules in the inner skeleton of a cell form a network of minute pipes and tubes, and this framework of biotensegrity allows the cell to move, reshape itself, or divide into two. At the cell surface, the tubules connect to other cells and form bridges, locking cells together and passing signals from cell to cell. There are tiny motors, kinesins, that move the tubules just as muscles move our joints. As we recognize shared patterns and scaled hierarchy, it may not surprise you to learn that these proteins, with their hinge-shaped structure, work quite the same way as actin protein molecules in our large muscle groups and those of other animals, both using the same hinge-shaped structure.

Cells are packed in tight, with contents under pressure, each inflated like a tire. Their contents are contained by the tension of the cell membranes, which don't just touch the membranes of other cells but bind together, holding one another tightly in place. Some are supported on a scaffold of basement membranes, which connect from cell to cell in a thin sheet like woven cotton. Other cells pack in tightly together as they grow and divide, squeezing and pushing one another, insinuating themselves into the crowd as best they can. If you want to picture the forces at work in a growing embryo, consider the tiny seedlings that push up through the concrete on the sidewalk; for each plant, as for cells, it is do or die. To live, each new growth faces the inescapable challenge to break through and erupt at the surface, and it is amazing to see how the tiny, soft leaf exerts enough force to push past massive obstacles in its path.

Cell growth and cell death

As cells in the human body divide and grow and divide again, the embryo grows in size and changes in shape. As growth progresses, some cells spread out into sheets while others roll into tubes. Also as part of this process, some cells differentiate into new cell types. Others self-destruct and die, to open a space between one area of growth and another, like those we saw at the end of the spinal cord, in Chapter 2.

To form our organs, limbs, and digits, the growing embryo balances cell growth and cell death. For example, until some cells between our fingers die back and open clefts, our early hands extend from our arms like paddles at the end of solid oars, our fingers fused together as one flat plate, webbed as in a duck or frog and unable to move apart from one another. Successful cell death then liberates the thumb and digits to move independently.[4]

Normal embryonic development calls for rapid cell division, then fast proliferation, differentiation, and migration of cells from one area to another. When a healthy biochemistry regulates and controls this process, growth proceeds apace and in predictable directions. But when something disturbs or disrupts those controls, the process goes awry, and significant deviation creates unwanted aberrations and malformation. The control systems that regulate activity and the pace of cell growth are similar for all stages throughout life. In utero, balanced chemical signaling yields a fully formed embryo, whereas dysregulation disorders growth and drives deformity. Later, in childhood and adult life, aberrant regulation of cell division, proliferation, differentiation, and migration invites tumor formation and cancer – disease reflecting an imbalance between cell growth and cell death.

Body plans

Looking at cars and how they have changed since their introduction about 150 years ago offers a concise visual parallel to the evolution of human bodies over the millennia. Consider the now-familiar four-door family sedan. Seventy years ago,

   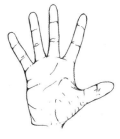

Figure 3.4

Development of the human hand. Cell death opens the clefts between the fingers and thumb to deliver full form and function.

few Americans owned an automobile – cars were quite dissimilar, each make and model distinctly recognizable. But with the building of highways in the wake of World War Two (a historical marker whose impact will emerge repeatedly as a touchstone in this book), conditions changed to favor automobiles as our primary means of transportation, and most US households now own a car. As automobile production has boomed, the bodies of the standard family car have become more and more alike, until today it is increasingly difficult to tell one make from another.

This change of auto bodies over time reflects an evolution because, to produce a vehicle for every family, today's car manufacturers all strive to fill the same goals for the standard four-door vehicle that so many buyers share: optimal space for comfort and safety, maximum storage, efficient fuel economy, and, therefore, aerodynamic contours. And symmetry-seeking teaches us that shared goals for function usually result in similar structures. Car companies have also been working within the same limits for construction, necessarily adapting to those shared constraints, such as maximum possible fuel economy within current technology, counterbalances between outer size and the space inside, how air flows around a moving vehicle, and costs of production. Competing for market share with other brands while juggling all these factors, most manufacturers now offer a more uniform product than the Studebaker, Oldsmobile, or Pontiac of yore – a sameness of shape, size, and features that is obvious at a glance.

As it is with cars over a single century, so it has been with animals over many millions of years. And evolution's effectiveness at designing animals for locomotion means that cars are notably similar to creatures. Like cars, every animal has a head end and a tail end. Everyone has a back and a front, a right side and a left side, just like the car with a front and a rear. The control center is in the front seat, with headlights like eyes, illuminating the world in front of you so that you can survey the territory and make decisions about your direction. Four wheels, like the four limbs of mammals, two at the front and two at the back, hold the body up off the ground and allow forward movement. And like both humans and other animals, the car has a front end that is more versatile, closer to the control center and able to direct changes, with air intake at the front, too, and waste exhausted from the rear. And these parallels between car and human are not just coincidence or metaphor: they're driven and defined by the necessities of locomotion. The symmetries such as we see in a car, we shall find everywhere in the human body.

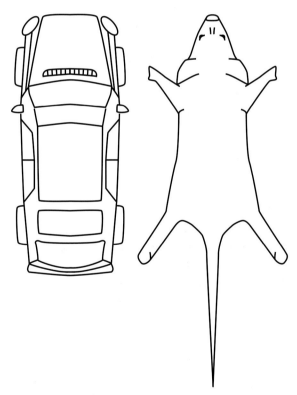

Figure 3.5

Automobile design and rodent body plan.

Back bones and Hox genes

Having only one way of making life and one body plan, symmetry squanders nothing in applying its one set of instructions for right and left, up and down, inside and out, concealing universal constants within outward variations and apparent diversity. Building the healthy body uses the same robust, well-tested codes over and over again, in frugal conservation, always the same toolkit with minimal waste and least expenditure. Everything matters, and each next step is dependent on the last, both for better and for worse.

Just as vehicle manufacturers use the same basic technologies to design both the family automobile and trucks or buses, nature shares the same instructions when designing all mammals, so they all share similar body plans. Each animal has a rod-like structure that stiffens an embryo's back, which will become the backbone around the spinal column. Each has a neural tube running along the body axis, expanding at the head end to form the brain, which will be protected by the bones of the skull. And each forms muscle groups as segmented blocks (called somites) that are distributed on the right and left sides of the spinal column. Each embryo grows a shoulder girdle and a pelvis, like the front and rear axles of a car. And each forms a gut tube and a heart on the underside of the spinal column, in a pattern quite the same for all mammalian embryos.

When a mammal builds limbs, obligatory economy means that it always uses the same set of instructions, the same genetic tools, over and over again for each limb, whether arm or leg, left or right, with a structure thus always similar though not quite the same. Under ideal circumstances, with an equal abundance of materials and optimal signaling and conditions for growth, all these parts are fully formed – more complete and symmetrical – but in

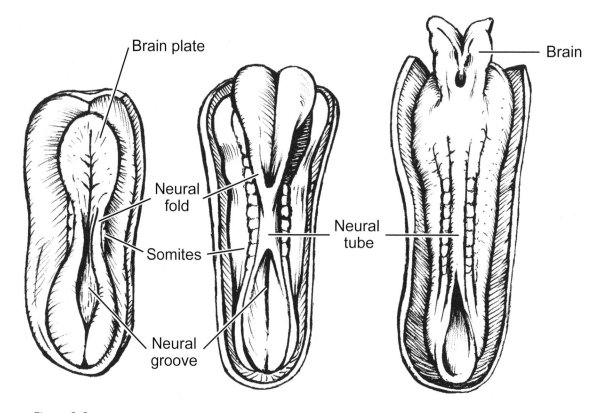

Figure 3.6

Human embryo somites.

less favorable conditions, the construction might fall short of the target, producing a less well-formed, less symmetrical final product. The impact of unfavorable conditions may land on only one aspect of development, but because of the shared rules and tools, one aberration is often matched by others, leaving patterns of deficit within the growing fetus – non-random associations in multiple areas.

The regulatory genes that define the body plan and drive its building are called Hox genes, and they hold ancient codes that have not changed for millennia. Their primary role is laying out the body plan of the growing embryo, head to tail. Together with just a handful of other ancient gene sequences, the Hox family of genes is responsible for directing construction of the whole body.

The body plan contains all of its assembly instructions in a single segment of the DNA code. Scientists estimate that there are about 20,000 protein-coding genes in a human being, but there are only 13 Hox genes, and they are responsible for the big picture – the foundational blueprint of the body. They're grouped together in a short, robust section of DNA, always arranged in the same sequence, to be read always in the same order, from head to tail. This fundamental segment is constant not only in all mammals but in reptiles, fish, and birds.[5] Any two humans share 99% of their genetic code with one another, and even more with our relatives. And we share much of our genetic code with other living creatures – 18% overlap with baker's yeast, 44% honey bee, 65% chicken and 85% cow. But we share the genetic code of Hox genes with innumerable species – all kinds of creatures.

This sameness is the reason our overall form is so similar to so many different animals. Hox genes define where the head end of the embryonic disc will be and where the tail, what will be the back of the embryo and what the front, and the position for the fore limbs and the hind limbs. The paired fins of fish, the wings of bats, the legs of a horse, and our own arms and hands are shaped by the very same master code using the same ancient toolkit. And one of their most important projects is laying down the nervous system, whose central shaft – the spinal cord – lies protected in the bones of the spinal column.

To build a body, the embryo must have clear assembly instructions – well written, indelible, and easy to read, like text. DNA is therefore a long string of genes that is read from left to right. 'String' is too flimsy a term to describe our genetic code for the body plan: it is more like a sturdy, old-fashioned metal zipper. When the slider runs down the parallel tracks of the zipper, it 'reads' the text of the DNA and opens a short segment into two. As the slider moves from one area to the next, it closes the zipper to leave the DNA restored and ready to be read again when another slider comes down the track later.

Genes direct all types of construction in the body, but Hox genes direct new construction, also producing the gene products that regulate other genes and trigger cascades of cellular activity. This ability defines them as the master genes, exerting decisive influence on the growing embryo.

In the embryo, our limbs grow out from the spinal column, with the upper limbs attaching at the sturdy bones of the shoulder girdle, mirrored by the lower limbs attaching at the sturdy bones of the pelvic girdle. And if we trace the back bones from one end through to the other, head to tail, we see that the size and shape of the vertebrae are different, grouping into distinct sections: small, lightweight cervical bones in the neck, transitioning to the rib-bearing thoracic bones that support the chest wall, and then the thick and heavy lumbar bones of the lower back, resting on the fused bones of the sacrum and finally our vestigial tail of coccyx at the bottom.

These transitions of spinal bones from one section to the next reflect distinct levels of Hox gene activity. Each level in the spinal column, from cervical at the top to sacrum and coccyx at the bottom, has its own corresponding set of regulatory genes that have shaped its vertebrae. Fewer are at work in the neck bones, more join in to help shape the thoracic bones, and more again are added at each new level, bringing Hox gene activity to regulate and direct how the sections are built. Reflecting nature's obligatory economy, not just the bones of the spinal column

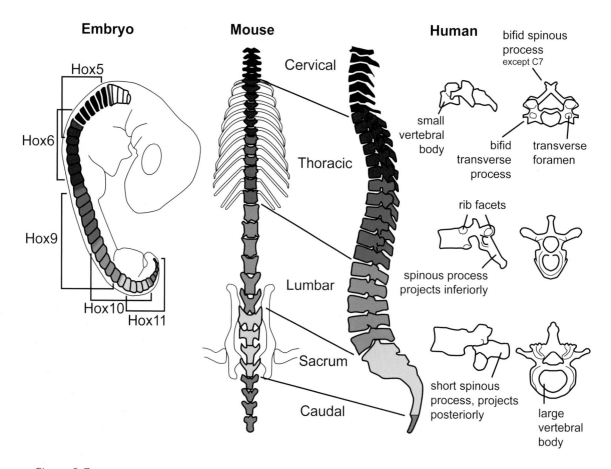

Figure 3.7

Spinal column. Left, mouse spines: shading distinguishes major types of vertebrae, from cervical above to caudal below and the corresponding Hox gene activation. Each major transition is marked by the addition of more Hox gene activity. Right, adult human anatomy demonstrates the physical characteristics of the vertebrae – the bones that make up the spinal column and surround and protect the spinal cord structures.

but all of the body's structures are defined by these levels. Orderly additions of regulatory gene activity transition from one level to the next to lay down our structures, and their success has consequences for function and, thus health.

Spinal segments are arranged stepwise, and they link to the limbs such that the nerves nearer to the top move the joints closer to the center of the body, while nerves further down move those further out towards the extremities. For the upper limbs,

then, the fifth cervical nerves control the muscles that move the shoulder, the sixth move the elbow, the seventh move the wrist, and the eighth control the long muscles that flex and extend the fingers; lastly, the nerves of the first thoracic segment innervate the intrinsic muscles of the hand that allow the fingers and thumb to work together. At each level, nerves that work to flex the joint work in concert with those that extend the joint, thus allowing optimal integration and linkage that leads to controlled muscle activity and smooth push-and-pull move-ment patterns.[6]

This stepwise linkage helps us understand spinal injury, such as might occur diving headfirst into shallow water. A violent strike to the head flexes and fractures the bones of the cervical spine at the base of the neck. This tragic accident will usually damage the spinal nerves and paralyze all the muscles of the body at and below the level of injury, so that the victim will lose all voluntary muscle activity in the lower body and the nerves that control hand and finger movement will be lost. But the muscles of the shoulder, elbow and wrist will typically be spared, because the nerves that move them lie higher on the spinal cord. These unfortunate patients are often active teenagers or young adults, and they can still move a wheelchair with their shoulders and arms but no longer move their fingers effectively for writing or feeding.

In the lower spinal cord the stepwise pattern is the same. Higher segments contain the nerve cell bodies that contract the muscles that move the joint closest to the body, the hip. Moving downwards, there are segments that move the knee and below that, nerves that move the muscles that move the ankle. The spinal segments that move the feet and toes are even lower down the stepladder, near the tip of the tail end of the spinal cord. At the head end of our central nervous system, where the spinal cord meets the brain, are the nerves to the face, and at the tail end are the nerves to the muscles of our pelvic floor. It is the pelvic floor muscles and fascia that provides the internal network of support that fills the floor of the bony basin that is the pelvic girdle, and they are critically important for bladder, bowel, and sexual function. For humans, there are no more spinal segments below the pelvic floor. As we saw in Chapter 2, the spinal cord peters out and ends here.

The embryo's grooves and tubes

To understand the patterns of the human body that Hox genes regulate – the whole body – it's essential to start with the embryo. There we can observe the central role of the spinal cord, especially the neural crest cells that emanate from the cells immediately adjacent to the cord, in shaping our initial conditions that form the platform for our health. The spinal cord provides the frame on which you can hang your growing knowledge of the body's function, just as the physical spine forms the axis from which an embryo builds a body, and the bones of the spinal column provide a trunk from which the body can hang all its limbs and organs.[7]

Tracing the spinal column back to its origins, long before an embryo lays down bones, we find its precursor appearing very early in embryonic development, when the mass of cells that will someday be an infant was only recently a blob. While the cells are still just a flat disc, a ribbon appears across it, which then becomes a groove. This line runs the length of the disc – just like the vertical diameter on Da Vinci's circle that encapsulates Vitruvian Man. We saw this disc in Chapter 2, when we considered the mermaid baby and its incomplete tail end, because this canal, which runs the length of the body, will run head to tail, or top to bottom.

If you reach your hand around to the center of your own back, you can feel the dorsal spines of your vertebrae, just under the skin. Your spine lands there because this canal, that appears early in embryonic development and is the precursor to the spinal column, forms along the back side of the embryonic disc. Following the symmetry-seeking principle to turn our object at different angles for observation, also imagine the disc as lying flat on a table: from this angle, the canal appears on the top surface. Think of other mammals who do not stand upright like humans but stay down on all fours, and you can easily see how the top is the back. Running a hand along your dog's back, you feel her backbone. To be consistent we call the back the 'dorsal' side, a term perhaps most familiar from the dorsal fins that stick straight up from the back of a fish – whose construction is also similar to humans', as Neil Shubin describes in *Your Inner Fish: A Journey into the 3.5-Billion-Year History of the Human Body*.[8]

This line running up the back is called the neural canal, centered on the disc and central to the embryo's task, for it will build the structures of a human body from this groove. The 'neural' of this canal means it will become the nervous system, and just as the brain will be supported and protected by the cranial bones of the skull, the spinal cord will be sheltered by the vertebral bodies and contained within the spinal canal – the passage through which the wiring of the brain communicates with all the rest of the body.[9] Within this canal will emerge the spinal cord and around it, its coverings

that will protect this vital set of structures. And on the underside another groove that will traverse the body from head to tail will form another vital system: the gut tube, and then all the organs and systems that grow from it.

The first step in all this building is to turn the neural canal into a neural tube – to fold up the sides of the groove and make a pipe. Because life is movement and flow, nature needs tubes to guide that flow and therefore builds them everywhere. Directed by the Hox genes, the embryo accomplishes this feat by raising the sides of the groove higher and

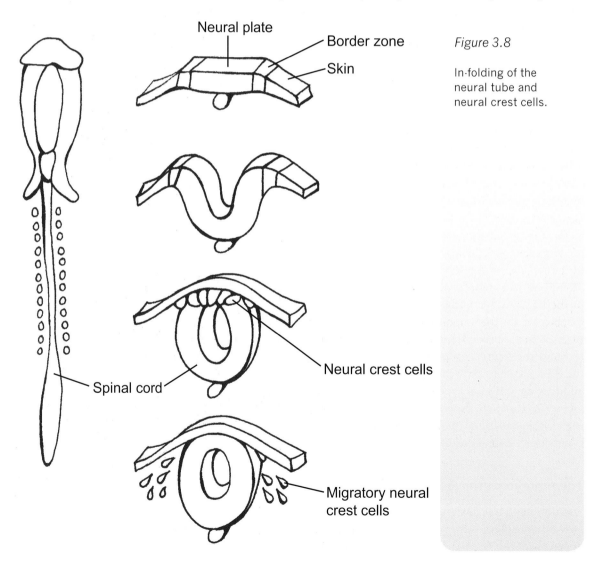

Neural plate

Border zone

Skin

Figure 3.8

In-folding of the neural tube and neural crest cells.

Neural crest cells

Spinal cord

Migratory neural crest cells

higher, like levee walls on either side of the canal. Special cells do the building, called 'neural crest cells', as they build the sides like cresting waves. Their rising on both sides deepens the groove, progressively more and more, until finally the edges fold together along the top, closing the seam to form a tube. A failure to fold adequately together and close the tube is the cause of spina bifida, a neural tube defect that, as we saw in Chapter 2, occurs most commonly at the tail end of the spinal column. Closure of the neural tube occurs during the first 6 weeks of pregnancy, before many women even know that they are pregnant.

Neural crest cells are busy during the early weeks of gestation, because they must lay down the template from which they can build our human structures. Working from their starting place alongside the spinal canal in the back, they also move forwards to the underside, to form another parallel tubal system: the gut. Where the dog's backbone is on the top and her belly is tucked away underneath, so our belly, which appears on the front of our body, forms on the underside of the embryonic disc. When we refer to our gut, we may mean only the stomach, but when doctors talk about 'the gut', they generally mean the whole digestive tract of organs, connected top to bottom – the gut tube. The neural crest cells form this tube in the same fashion that they form the neural tube: building up the sides then rolling in the edges and zipping the gap closed, to form a tube. Given the importance of digestion – of nourishing the growing organism – the gut tube is among the earliest structures to form in the new embryo, right alongside the nervous system. The pioneers who first studied the building of a gut tube, by viewing chick embryos with a microscope, named the process 'gastrulation', literally 'the making of the stomach'.

Like both fish and chickens, all mammalian embryos begin their growth by making a neural tube along the back and a gut tube along the front, or underside. The relative success of the neural crest in building a good tube early on remains apparent throughout life, because the neural crest supplies

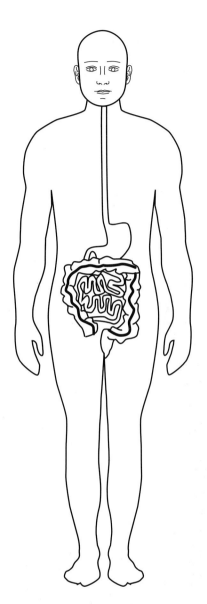

Figure 3.9

The gut tube passes like a tunnel from mouth to anus, and each transition in form reflects the addition of more Hox gene activity.

all the nerves that populate the full length of the gut. These nerves drive effective bowel motility, so if your bowel movements are easy and elimination habits are dependably regular, you can thank your neural crest cells.

While we have multiple digestive organs, they comprise just one gut tube. Think of this tube like a tunnel. The gastrointestinal tract runs from the top of our bodies to the bottom, opening to the outside at the mouth and passing through to open again to the outside at the anus. The gut walls are lined with an inner surface which, though moist, is otherwise like the dry outer skin that covers our outsides. This tissue lining the gut tube also serves the same purpose as the skin on the outside of our body, as both barrier against and interface with the external environment. At either end, the openings reveal the junction zones where the dry skin of the outer body transitions to the moist lining of the gut tube, at our mouth and anus, respectively.

When you swallow food, it doesn't pass fully into the body so much as it passes through the body, while staying within its tube. Neither brownies nor bell peppers can detour directly to the heart or to a hand, as they must stay contained and travel along the main route. The face – our mouth – is the one entrance for this open-ended tube, and the anus is its one exit. The gut traverses the whole body, carrying ingested food from the outside back to the outside, allowing only certain microscopic elements of the food to filter through the walls into the bloodstream and thus on to the rest of our bodies. The remaining indigestibles emerge at the anus, as waste. Mouth and anus are thus complementary opposites, of entrance and exit.

Tops and bottoms: shifting our gaze to the dorsal view

Da Vinci's man in a circle and square is naked, fully illustrating the reflective symmetries of upper and lower body. This familiar artistic nude may not prompt us to consider the parallels between our sociable, sharable faces and our well-covered private parts, but symmetry-seeking challenges us to do just that: to compare how faces and posteriors might be similar to one another, on the same body or across two bodies. But when we look at the digestive system

for top-bottom symmetries, it seems to offer instead an asymmetrical mess.

When medical students explore internal anatomy through dissection, they start from the front, opening a body's soft underbelly, from the breast bone to the pubis, and from this point of view, our human organs look like a hodge-podge collection with no obvious logic to their placement. In a normally developed body, the spleen is reliably on the left, balanced by the liver and gall bladder on the right, but the gastrointestinal tract lies coiled like a great snake inside the abdominal cavity, with no visible orderliness to its form – a strange contrast to the tidy left-right, top-bottom symmetries of the body's outer limbs.

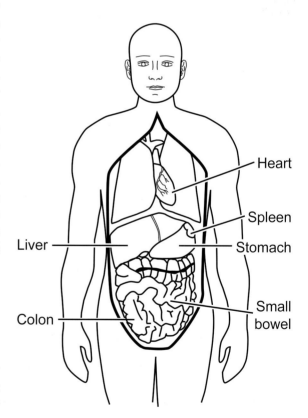

Liver
Heart
Spleen
Stomach
Colon
Small bowel

Figure 3.10

The abdomen, part opened to reveal the gut tube and organs from the front.

Just as the logical simplicity of DNA's double helix was hidden for so long within the tangled, twisted ropes of chromosomes in the cell nucleus, the actual straightforward simplicity of the gut can remain hidden even when the body's internal cavity is laid open to the human eye. But if we come to understand the digestive system as one long tube and study its development and structures as a whole, it provides a model and template for understanding all tubes in the body, both their form and their function. Each tiny kidney tubule, for instance, is a miniature replica of the giant-sized gastrointestinal tract.

Developmentally, the gut tube follows the same pattern of Hox gene regulation as the spinal column, increasing the number of Hox genes involved as you move down from head to tail. Accordingly, the esophagus uses only the first three Hox genes, and the transition to stomach adds two more (five in total) and then the duodenum has six. Each new segment of the gut tube where the shape and form, structure and function transition from organ to organ shows a change in the length of the sequence, with each step requiring the addition of more regulatory genes. Finally the rectum, at the end, uses all thirteen Hox genes. And for the organs that grow out from the gut tube, such as the liver and pancreas, lungs, or kidneys, each starts over with a new beginning, running through the Hox gene

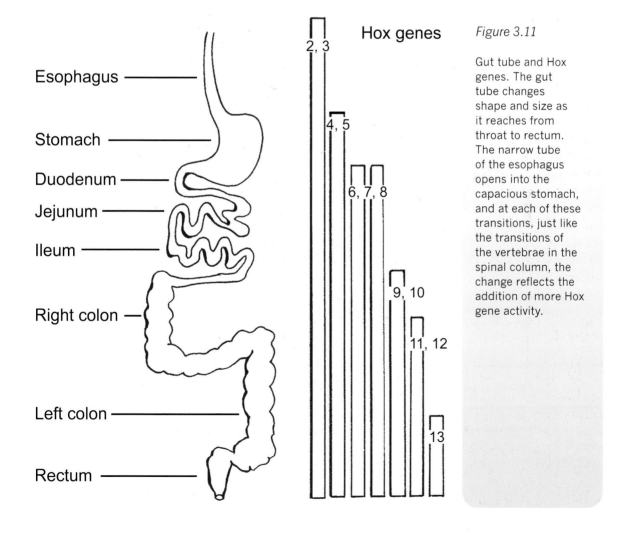

Hox genes

Figure 3.11

Gut tube and Hox genes. The gut tube changes shape and size as it reaches from throat to rectum. The narrow tube of the esophagus opens into the capacious stomach, and at each of these transitions, just like the transitions of the vertebrae in the spinal column, the change reflects the addition of more Hox gene activity.

Esophagus

Stomach

Duodenum

Jejunum

Ileum

Right colon

Left colon

Rectum

2, 3

4, 5

6, 7, 8

9, 10

11, 12

13

sequence adding more activity at each step, until each new organ system is fully built.

Like the rest of us, practicing physicians also look at the body mostly from the front, face-to-face with their patients. But symmetry-seeking invites us to consider alternative views, and when we survey the organs inside the body from the back, the orderly pattern of the internal body plan becomes clearer. If we also bring our understanding of the gut as a tubal system – similar in its development to the neural tube – with all its organs simply waystations on one long tube, its profound symmetries emerge. And to fully comprehend the digestive system, we must also consider the organs of other systems closely linked to the gut, systems that grew out of the gut tube in the embryo. The kidneys and the lungs, especially, help to illuminate the body's top-to-bottom internal symmetry.

Imagine, then, a tall window in the back of the human torso, stretching from the neck all the way down to the pelvis. Look past the bones as if they were transparent, to the soft tissues of the organs. Through this dorsal window, we can see a clear division between the chest cavity above and the abdominal cavity below, with the diaphragm separating them. If you are looking for similarities above and below, you notice the paired lungs in the chest and the paired kidneys in the abdomen. After we have spotted this similarity, symmetry-seeking prompts us to apply its other principles to these two paired organs, the lungs and the kidneys, to see what we can learn.

Lungs and kidneys

In the embryo, both the lungs and the kidneys first arise as outgrowths from the early gut tube. Well before these two mirrored organ pairs evolve into systems separate from digestion – the respiratory and urinary systems – the buds of each of them appear at exactly the same moment in embryonic development, the 28th day. Then Hox genes drive them to branch out by cell division – building tubules, that favored structure in nature – continuing in parallel growth through eighteen levels of branching divisions, until the lungs and kidneys are complete.

Modern medicine treats each of these organs systems with a separate specialist – what could be more different than the lungs and the kidneys? The lungs are air-filled bellows in the chest, while kidneys are solid bean-shaped organs that filter the blood to clean it and make urine. Pulmonologists and anesthesiologists would have no hesitation in declaring that there is no other organ like the lung, while the nephrologists or my fellow urologists have no doubt that there is truly none like the kidney. But the structure and functions of these two systems, developed in parallel, remain closely linked, both in embryos and infants, and throughout

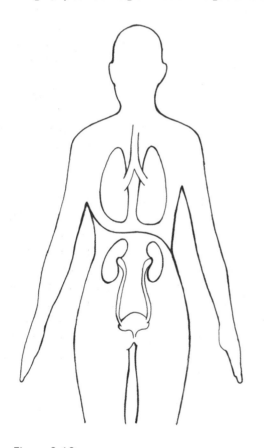

Figure 3.12

Dorsal (back) view of chest and abdomen.

human life. Both are systems of excretion, on which we depend heavily. The human body cannot tolerate an obstruction of flow in either for long, before progressing towards death – by uremia, if the kidneys are blocked, or by asphyxia, if the lungs are obstructed.

We recognize kidneys as organs of excretion, sending urine out through the bladder, but we tend to think of the lungs' primary job as the intake of oxygen. In reality, the lungs' more pressing and essential task is ridding the body of excess carbon dioxide (CO_2). We can recognize the symmetry of their roles in excretion when we remember that a state trooper can test for blood alcohol levels from either end of the body: in urine or on the breath. Traditionally, physicians made sure to note the smell of the breath as a deliberate part of the bedside assessment of an ill patient, during a classical physical examination that considered the whole body. In some diseases, the breath carries distinctive smells excreted in exhalation: in a diabetic coma, the odor is often sweet and flowery, but liver failure smells sour and bitter from ammonia, and kidney failure is stale with urea.

A pioneer of modern medicine spotted this parallel between lungs and kidneys 150 years ago. While Charles Darwin was working on *On the Origin of Species in England* (1859), Claude Bernard was in France writing his classic work *An Introduction to the Study of Experimental Medicine* (1865). Bernard set the cornerstone of the science of physiology, and he laid out the principles of the scientific method of research. He was the first to consider how the near-constant conditions of the internal environment of the human body are regulated and held in balance, and he recognized that the organs of excretion included not only the kidneys but also the lungs and even the skin. As Bernard wrote: 'the products of organic decomposition are collected in the blood and circulate with it to be excreted, either in the form of gas through the lungs or in the form of liquid by the kidneys'.[10] The kidneys waste nitrogen from the body as urea, conserving water, while the lung wastes carbon dioxide and also conserves water. It happens that in the process of excreting carbon dioxide, the lungs imbibe oxygen.

Seen with a symmetry-seeking eye, lungs and kidneys show strikingly similar structural patterns and developmental landmarks. In the lung, branching buds grow into a tree-like pattern, and the air sacs are clustered like leaves on the smallest twigs. In the kidney, the buds bifurcate and branch in the same way, but the fluid-filled spaces coalesce to form larger funnel-shaped channels for collecting and directing the urine flow. Fluid production in the early kidney may contribute to the forces that shape the urinary tract. By the time of birth, kidneys have been filling with fluid and making urine for the last months of pregnancy already, using the urinary tract and shaping it at the same time. If a kidney is blocked because of a congenital anomaly, it might blow up like a child's party balloon, which then shows up as a fluid-filled abnormality visible on fetal ultrasound scanning.

The tiny tubules of the lung and the kidney are like a labyrinth of tunnels in a mine, and they both face the same challenge of how to hold open the flimsy channels and prevent collapse. A mine needs support beams and lagging, but nature uses a simpler solution, delicate and elegant: the lining cells secrete a special surfactant (phospholipids) that forms a thin layer, like the surface of soap bubbles. Surfactants hate water, and they spread out and convert the surfaces to be an impenetrable duck's back, whose key functions are to hold open the tubules and prevent collapse, keeping the surface clean and passages clear. Infants who lack one of the genes to make these compounds suffer severe respiratory distress, and even minor flaws in this function foretell both lung and kidney problems in later life.

In both the lung and the kidney, there are about a million tiny exchange membranes that constantly filter the blood at a capillary bed, in the alveolus and glomerulus respectively. The waste product is displaced into the excretory tubules that unite to form ever-larger tubes for carrying waste out of the body. After birth, each breath blows off unwanted carbon dioxide, at the same time drawing in fresh air and oxygen in exchange, many times a minute. Urine is stored temporarily inside the body, then voided at convenient intervals about 5 or 6 times a day.

The link between these two mirror systems appears early, as the lungs and the kidneys tend to reflect each other in multiple ways. If there is a failure in one, physicians know to look for a failure in the other, because abnormal patterns of development in the respiratory system typically mean abnormal development in the urinary system. If, in the early work of neural crest cells, a bronchial bud or ureteric bud fails to grow, the newborn will be a missing a lung or kidney, and other anomalies tend to co-occur in the same individuals as well, especially when there is duplication, such as a double lung mirrored by a double kidney, or cystic diseases in both systems. Problems with mucus plugs in the airways, such as one finds in cystic fibrosis patients, are often accompanied by stones in the kidneys. All these parallels help us to understand how much these symmetrical systems are linked in any one person's body, in healthy patients as well.

As organ systems that first appear by branching from the gut tube, the structures for breathing and peeing also remain physically close to the digestive tube, from top to bottom, and functionally linked to it, as well as to each other. For example, when a newborn's gut has failed to develop properly, with the tube failing to close fully, the esophagus can be left open into the airway. Babies with this tragic condition turn blue on first feeding, as milk fills their lungs instead of their stomach. Surgical repair of this anomaly must usually be followed by a second surgery, because the same newborns may have a parallel problem at the other end of their bodies: a blind-ending (closed off) rectum. Until it is opened to the outside, their digestive system diverts into a parallel tube, such as the urethra – which is the kidneys' mirror to the lungs' airway.

In some cases, a blind-ending rectum leaves waste to divert into the vagina, another neighboring tube, which is part of another system that also grows out from the gut, the reproductive organs. And as we consider top-bottom symmetries in the human body, sex organs and their analogues in other systems can teach us a great deal.

Exposed faces and covered places, upside and down

Patterns of top-bottom mirroring reveal how closely our heads and tails are linked. Newborns with abnormal kidneys and lungs also characteristically have abnormal facial appearances and, at the other end, anomalies of their genitalia. Other links in developmental problems appear on and near the body's surface as well, such as a mirroring between the penis and the nose. When the skin fails to close properly in either of these organs, the most common pattern of structural defect is the same: a cleft in the bridge of soft tissues that lies between the gut tube and the respiratory or urinary tract. On the face, at the respiratory end of the body, the resulting malformation is a cleft lip and palate, and at, the urinary end, the result is a split opening at the underside of the urethra, called hypospadias. These defects, one at the top and one at the bottom, are usually found independently, but they are structurally the same, and there is a statistically significant concurrence of these two types of defects in the same male newborns.[11] Such mirrors remind us of the instructive links between the top and bottom of our bodies, and our knowledge of one end can help us better understand the other.

The reproductive system is one the three tubal systems that traverses the pelvic floor, along with the digestive system and the urinary tract. In both females and males, the sex organs bud from the same tube systems that form the kidneys, which, as we have seen, emerge from the gut tube. Because reproduction is so critical to the species' survival, growing the reproductive organs is a developmental priority, so they appear early in the embryo – at six weeks – only two weeks after the tube system that gives rise to the kidneys from which they branch out. The joining of egg and sperm performs a necessary mixing and remixing for many of our genes, introducing variations for new beings, even while some genes, like the ancient sequence of Hox genes, transfer quite untouched and untouchable, for they alone hold the basic and critical codes utterly essential for building new life.

In the same way that the top and bottom mirror one another in the lungs and kidneys, facial features and the genitalia share reflective symmetry during sexual function. The face blushes and the cheeks, lips, and nose flush during arousal and sexual readiness, mirroring vascular excitement in the genitals, so that the outward facial appearance resembles the perineum. And because nature is basically simple and relentlessly economical, no tool or mechanism is used only once for a unique singular task, so we find structures highly similar to the sexual organs at the top end of the body, in the respiratory system. Even orgasm has a parallel upper body event: the sneeze.

As for sexual climax, the trigger for a sneeze can be a light, repetitive touch to nearby membranes, such as a tickle with a feather. The stimulation leads to a step-wise change in breathing pattern, with a deep, involuntary drawing of the breath just before the culmination of the sequence. The vascular tissues of the nose in preparation for a sneeze – as in the genitals with orgasm – engorge and increase their secretions. And the climactic moment is often preceded with an aura and usually involves the whole body and a forceful expulsion, focusing sensory and motor activity to the exclusion of all other tasks. The eyes close, and the whole-body posture can change, almost like a convulsion, in a pattern sometimes repeated again and again, through a series of sneezes. Thoughts suspend, and the sneezer may utter and repeat involuntary sounds, including whimpers, while episodic fluid discharges are emitted. The expulsive forces that drive the emission come from many muscle groups working together to produce coordinated contractions while other muscle groups are simultaneously inhibited, as the sneeze momentarily interrupts all other muscular activities. For example, the diaphragm is completely relaxed at the moment of emission – the reason that sneezing might bring a bothersome risk of accompanying urinary leakage. But at the moment of the sneeze, involuntary functions prevail in the body. Then afterwards, the sneezer experiences relief, tranquility, and calm.

With such a parallel experience between the genitals and the nose during sex and sneezing, plus their shared response to sexual arousal, it may come as no surprise that the nose and the penis also rise and fall together in response to medications for erectile dysfunction. When drugs are used to treat erectile dysfunction, they increase blood flow and enhance erection, but they also light up your complexion so noticeably that your partner can easily guess that you have taken a pill, because of your red face. Genital skin and facial skin mirror one another, so as penile vessels dilate, your face becomes slightly puffy. Furthermore, reminiscent of the sneeze, these same drugs can interfere with breathing for many men, because it gives them a stuffy nose. The lower urinary tract is lined by vascular spaces fused to bone, and so is the upper respiratory tract, where sinus cavities and nasal spaces act as air conditioners to warm and moisten the inspired air on its way to the lungs. Engorgement of these spaces often brings on a feeling of congestion, as the reproductive system and the urinary system share one tube in men, in the penis. As we've seen in comparing kidneys and lungs, the urinary system reflects the respiratory system, so it makes sense that a drug given to promote penile erection might also give the man a stuffy nose.

Symmetry of thyroid and ovary

Not only whole organ systems but specific glands mirror one another between the top and bottom of the body. Mirroring the reproductive glands found in the lower bodies of both men and women is the thyroid gland in the upper body, and these two systems have a design and organization similar to one another. Just as the reproductive glands emerge from the tube system that forms the kidneys, the thyroid arises adjacent to tube systems close to the airway, and – as you might guess by

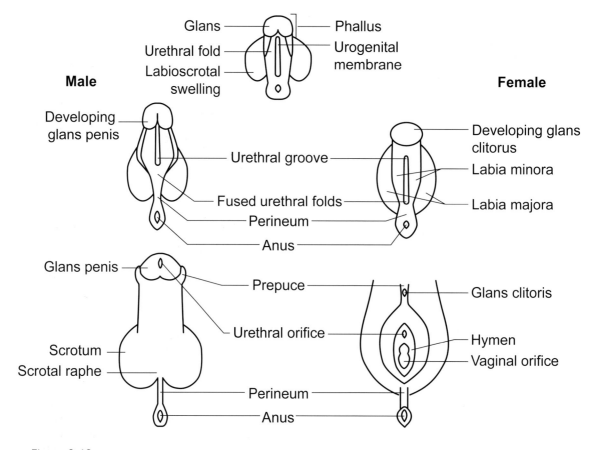

Figure 3.13

Genital anatomy reflects the shared origins of all parts, penis/clitoris, scrotum/labial folds. Sexual development depicts change of emphasis – all structures are represented in males and females, but some amplified more than others.

now – at the same developmental moment. They remain near to those systems, so that the thyroid is immediately adjacent to the voice box (larynx), which is the gateway between the atmospheric space and the internal spaces of the respiratory system and, in the same way, the ovaries and testicles are adjacent to the lower urinary tract and the urethra, the gateway between the internal space of the urinary bladder and the outside.

In both systems, there are two similar parts: one on the left, one on the right, bridged across the midline by connecting structures. The bridge of the thyroid – the isthmus – is structurally simple, while the tube structures for reproduction are more lengthy and complex, but as we use symmetry seeking, this difference ceases to disguise their broader similarities. In its place just proximal to the larynx, the thyroid surrounds the airway, and the cervix or prostate abuts the bladder base and urethra.

Not only are the structure and design of these systems symmetrical, but there is typically great harmony in the way they rise and fall together. Both the thyroid and the gonads promote growth and development, and the pituitary hormones that stimulate their activity are almost identical in structure – thyroid-stimulating hormone (TSH)

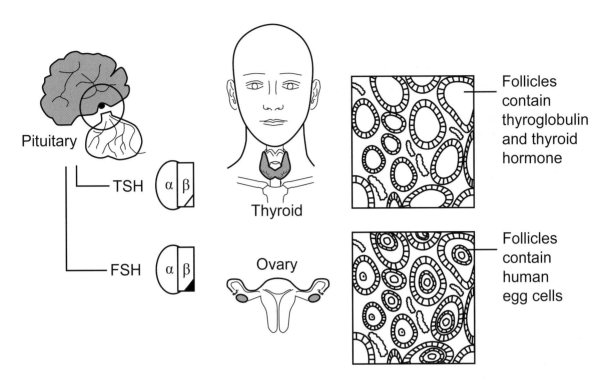

Figure 3.14

The anterior pituitary gland releases several specific stimulating hormones to regulate cell activities including growth and development, and to maintain a stable internal balance throughout the body. Thyroid-stimulating hormone (TSH) and follicle-stimulating hormone (FSH) are glycoproteins that stimulate the thyroid and ovaries respectively. Each is made of two chains of amino acids – the alpha chain is identical for both and the beta chain is only slightly different.

Thyroid and ovary are very similar in structure. Both include vascular and connective tissue elements supporting nests of cuboidal cells arranged in oval shaped follicles of different sizes. In the thyroid, follicles are filled with acellular colloid-rich stores of thyroid hormones, but ovarian follicles contain both colloids and egg cells. Small follicles contain immature egg cells and the larger ones hold more mature eggs ready for release with future rounds of ovulation.

and follicle-stimulating hormone (FSH). The parts of the brain that monitor and respond to levels of hormone in the blood (the hypothalamus and the pituitary) drive the release of both these stimulating hormones. (We'll look more closely at these brain structures in Chapter 9, on the brain.) And like TSH and most hormones, FSH is the same in men and for women. When released into the bloodstream, FSH energizes either the testicles in men or the ovaries in women, and we see mirror effects in the thyroid.

The thyroid and ovary are reciprocal partners and, as always, the harmony of hormone balance is delicate yet dynamic, designed to adjust and reset from moment to moment to meet the continually changing needs of a body's current conditions. Parallel events in a woman's reproductive system and thyroid reveal the linkage between these top-and-bottom glands, especially with pregnancy and during menopause. A clinical sign of pregnancy is swelling of the neck, recognized since ancient times, when Egyptians tied a reed around a young bride's neck, waiting for it to break. What causes the neck swelling in early pregnancy is the thyroid. Human chorionic gonadotropin, a hormone made by the placenta, is a stimulating hormone not unlike TSH and FSH. As it stimulates the ovaries to perform their tasks for pregnancy, it also stimulates the thyroid, which responds by growing larger.

Reflecting the symmetry of pattern between ovaries and thyroid is menopause. Where pregnancy increases activity in both systems, the end of the reproductive years triggers a decrease in both, as monthly ovulation trails off and levels of female sex hormone fall. The most common endocrine problem that appears for some women during menopause is slowing of the thyroid gland. Slow and insidious, hypothyroidism linked to menopause tends to creep up slowly, because the hormone levels drift off quietly over time, slowing metabolism and even changing appearance.

Symptoms of menopause and hypothyroidism are related and overlap, in particular weight gain and fatigue, sensitivity to cold, slowing thought and speech, dryness and thinning of hair, and changes of voice and skin. Because many women and their doctors may assume that these symptoms are all effects of menopause itself or simply of aging, the slowing thyroid – a distinct but easily treatable problem – can go unrecognized and untreated.

Symmetry of biorhythms

The nervous system uses electrical signaling to control muscle and movement, but it uses chemical messages to direct and regulate body functions. The electrical signals travel along nerves (in the neurological system), and the chemical messengers are hormones, the internal secretions managed by the endocrine system. Hormones are molecules potent for signaling and switching, made and secreted by specific endocrine glands, such as ovaries, testes and thyroid, and also including many others, such as the pituitary, the adrenal glands, and the pancreas. Most of the body's glands make fluid secretions, such as sweat or saliva, but endocrine glands are ductless,

and their job is to make hormones that are released directly into the blood, to travel throughout the body and act on particular targets.

Together, nerves and hormones act in harmony as the neuroendocrine system. The nerve cells of the hypothalamus and the pituitary, both in the brain, communicate closely with one another and regulate the hormone glands and tissue functions in the rest of the body. Some hormones are released in tiny regular pulses, like a ticking metronome, but the release of most hormones is not at all constant. The body monitors activity and steers the course, like a driver responding to traffic – sometimes speeding, sometimes slowing, always making tiny incremental corrections – and biorhythms reflect the changing cycle of oscillations between more activity and less.

We will revisit hormones in the next chapter, to consider other factors – some of them genetic, some internal and some external, some manageable and others less so – that threaten to disrupt balance and disturb the healthy steady state.

Symmetry would predict that any medication strong enough to have one effect in the body will always have more, because cells, tissues, and organs are all made from the same materials and share so much. Marketing is targeted and advertisements imply that a prescribed medicine will go straight to the problem, like a smart missile finding a bull's-eye, but nothing could be further from the truth. As we pursue symmetry in the body, we will appreciate that medication, once in the body, will go everywhere it can and influence every possible cell that is susceptible to its action.

References

1. Wolfe SW, Hotchkiss RN, Pederson WC, et al. (Eds) (2011). Green's Operative Hand Surgery, 6th edition. Philadelphia: Elsevier. Chapter 54.
2. Klucevsek K (2013). The Cytoskeleton: Microtubules and Microfilaments. Lesson 9 of Chapter 6 – Cell Biology: Tutoring Solution. Online: https://study.com/academy/lesson/the-cytoskeleton-microtubules-and-microfilaments.html#transcriptHeader.
3. Ingber DE (2006). Mechanical control of tissue morphogenesis during embryological development. *International Journal of Developmental Biology*, 50: 255–266.

4. Zeller R, Lopez-Rios J, Zuniga A (2009). Vertebrate limb bud development: moving towards integrative analysis of organogenesis. *Nature Reviews. Genetics*, 10: 845–858.

5. Fischer BE, Wasbrough E, Meadows LA, et al. (2012) Conserved properties of *Drosophila* and human spermatozoal mRNA repertoires. *Proceedings of the Royal Society B: Biological Sciences*, 279(1738):2636–2644.

6. Skaper SD, Moore SE, Walsh FS (2001). Cell signaling cascades regulating neuronal growth-promoting and inhibitory cues. *Progress in Neurobiology*, 65: 593–608.

7. Tsiaras A (Nov 2011). Conception to birth - visualized. Online: https://www.ted.com/talks/alexander_tsiaras_conception_to_birth_visualized.

8. Shubin N (2009). Your Inner Fish: A Journey Into the 3.5-Billion-Year History of the Human Body. New York: Vintage Books.

9. Philippidou P, Dasen JS (2013). Hox genes: choreographers in neural development, architects of circuit organization. *Neuron*, 80: 12–34.

10. Bernard C (1865). Introduction à l'étude de la médecine expérimentale. Paris: J.B. Baillière et fils.

11. Fernandez N, Escobar R, Zarante I (2016). Craniofacial anomalies associated with hypospadias. Description of a hospital based population in South America. *International Brazilian Journal of Urology*, 42: 793–797.

Hormones, the common currency across species

Disturb an angry dog, and an outpouring of hormones instantly triggers its inner alarm: a sudden rush of adrenaline causes its heart to pound faster and muscles to tense. This surge from the adrenal glands immediately brings its whole body to attention, arousing the brain, quickening the breath, and dilating the pupils, as adrenaline courses through the bloodstream, triggering a cascade of effects in preparation for a fight or a flight. This reaction occurs in dogs and humans alike. Both release these identical adrenaline molecules when triggered by sudden fear and stress, discharging this primordial chemical into the bloodstream by reflex. In both humans and dogs, blood vessels carry the adrenaline to all corners of the body, instantly sounding the alarm, readying the organs and all of the body parts to face an immediate challenge.

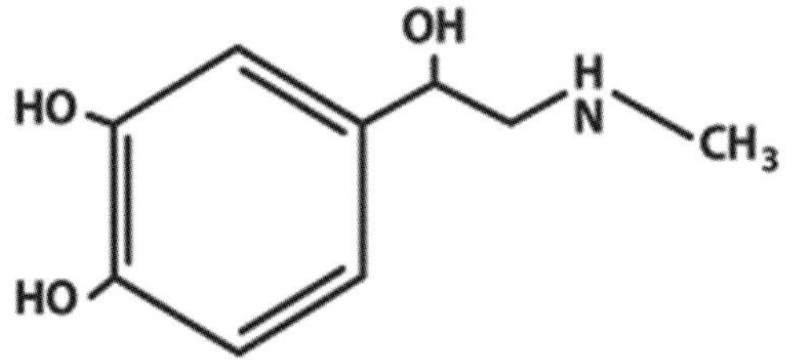

Figure 4.1

The adrenaline molecule has a simple structure of one ring of six carbon atoms and a short tail.

Adrenaline is one hormone of many – all primal chemical messengers shared by every mammal. Perhaps the most familiar hormones are those associated with sexual development, as they cause visible differences in females and males. During puberty, sex hormones cause differences to emerge between the sexes, with testosterone in adolescent boys directing them towards broader shoulders, a narrow pelvis, and coarse new facial hair, while estrogen leads girls to narrower shoulders, a wider pelvis, and breasts budding. These same hormones prompt novel appetites and altered behaviors, too, as they set bodies on a new course. And this universal currency of hormones, which directs growth and development, shapes physical features and drives action and emotion, we share not only with dogs, cats, and laboratory animals, but with animals in the human food chain, as well as with some plants. So there are multiple opportunities for these potent chemical actors to enter human food and thus the human body, where they can trump our own bodies' instructions and redirect human cell biology to their purposes.

Universal platform

The origins of the animal body plan provide a clue to our awesome ancient heritage and the shared platform of embryonic growth, from which springs all contemporary mammalian construction, for animals and human beings alike. The foundation of the body plan is a template, defined by the genetic code common to all mammals, but also driven by energy flow and shared physical laws.

The Hox genes we saw in Chapter 3 regulate development, along with other master genes, and one of the hallmarks of the animal body plan is bilateral symmetry. Our bodies cannot conceal these inborn patterns, imprinted as they are on the outside of the body, having appeared early on in an embryo's many symmetries.

It is fantastic to imagine that all the body's necessary know-how to build a full adult form for development, growth, and ongoing lifetime maintenance is packed into one single cell, the tiny fertilized egg. Like a capsule on a mission to the moon, all the instructions and tooling required to get the job done are packed into this one cell, where competition for vital space sets severe limits and dictates strict obligatory economy.

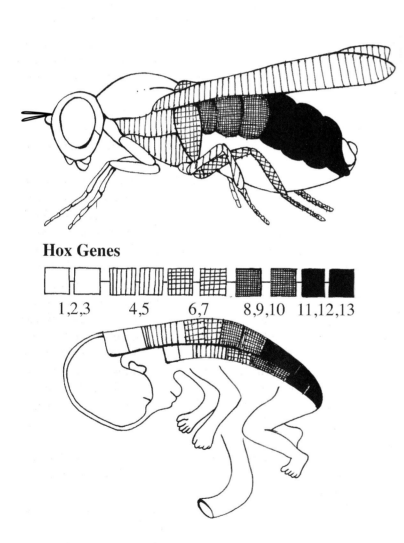

Hox Genes

1,2,3   4,5   6,7   8,9,10   11,12,13

Figure 4.2

The fruit fly (*Drosophila*) has two clusters of homeobox or Hox genes, each having eight genes. These ancestral genes determine the characteristic features of the fly's body from head to tail, positioning each body part at the correct site. Mutations in these genes lead to formation of body parts at the wrong site, such as a leg where an antenna should be.

Our human Hox gene sequences consist of a cluster of 13 regulatory genes always transcribed in the same order, and, as in the fruit fly, they have the same critical role of orienting the body plan from head to tail. Every human cell has four discrete copies of Hox genes, each securely housed on a different chromosome. Now demonstrated in all animals, from the simplest worms to the largest mammals, Hox genes provide evidence that the most complex of animals share some common features with the simplest of life forms, suggesting shared ancestral roots.

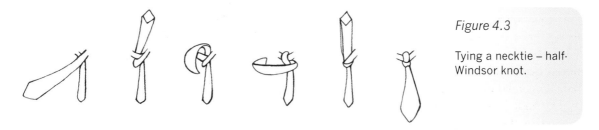

Figure 4.3

Tying a necktie – half-Windsor knot.

Over the great span of evolution, the body chooses simple rather than complex, and only the most reliable, tried-and-true systems prevail. Like a chef who prefers versatile tools in the kitchen, genes favor universal tools that can perform multiple functions, rather than unique or specialty elements with singular, transient, or non-essential roles. One sharp knife and a sturdy bowl, each with any number of uses, serve better than specialized implements with only one use. This necessary versatility constrains biological systems to conformity, repetition, reiteration, and, therefore, symmetry – some of it displayed openly, with other instances more obscure.

The complicated task of building a body has something in common with the simple tying of a necktie: the tendency to find a familiar process that works well with as little trouble as possible.

Now confirmed by mathematical proofs and computer models, it turns out that there are exactly 85 possible ways to knot a piece of cloth around the neck. Enumerating all the possibilities in *The 85 Ways to Tie a Tie* and describing the current favorites, physicists Thomas Fink and Yong Mao reveal that out of all these possible choices, only one or two ways are favored.[1] One knot, the 'half-Windsor' (named after its origins in the British royal household), has proven enduringly popular due to its economy: it offers less work than most knots, with fewer turns, a symmetrical shape, and the ease of self-releasing. Just as, out of the many possible choices, tie wearers choose only a few to use again and again, favoring the quick and easy over the time-consuming and elaborate, so it is in nature. Among cells, hormones are no exception: like the ties that are mostly knotted the same, hormones have evolved with stunning economy to be structurally and functionally similar across species.

Consider the growth of a limb, similar in all embryos – the legs of a horse, the wings of a bat, or the arms and hands of a human, all following the same pattern. As we saw in Chapter 3, once nature has found a method of limb-making, it revises and repeats it over and over in all different species, with the exact same sequence of genetic signals at work but only minor modifications of timing or spatial responsiveness to shape specific forms and functions.

As they bud in their precise locations, limbs in these varying species are responding to the same chemical signaling and cascades of gene products that regulate growth and have done so for eons. Even between limbs destined for very different purposes, the similarities are much more marked than the differences, once we are looking for similarities. As Adrian Bejan explores in his chapter 'Animals on the Move' in his book *Design in Nature*, the different iterations of animal limbs preserve the fundamental design across species, changing only the proportions and emphasis.[2] So even as animal life exhibits its extraordinary outward diversity, beneath it all the basic template is the same, reducing some elements and amplifying others, but following the same essential pattern for all.

Symmetry-seeking helps to reveal similarities and explore shared features. Humans have an outward appearance like no other animal, but as we explore human features using symmetry-seeking, we find that their basic template is still the same. Some elements are more prominent than in other animals and others less so, but the resulting patterns are invariable. Like other

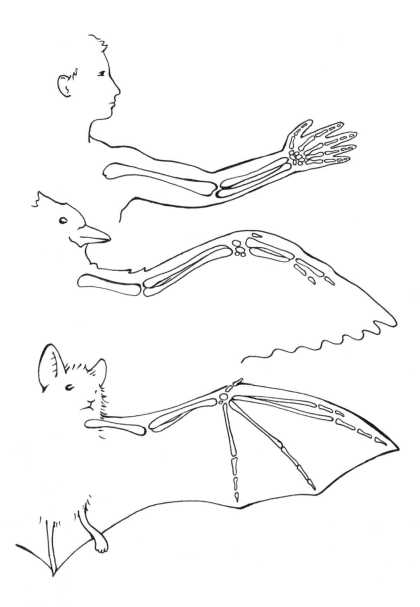

Figure 4.4

Homologous structure of fore limbs: the same Hox gene signals are at work across all species, so the pattern of limb development is very similar – whether in growing bat or bird wings, or the human arm and hand.

mammals, we breathe air, drink fluids, eat food, urinate and defecate, and – on our insides as on our outsides – we share structures and systems for basic life tasks, evolved over eons.

For all mammals, the ticking metronome of the hypothalamus releases tiny regular pulses to sustain biorhythms, monitoring activity and oscillations, sometimes speeding or slowing to make tiny corrections moment to moment, but always cycling, steering the course.

For all mammals, the hypothalamus (a section of the brain) and a gland closely linked to it, the pituitary, together regulate the other hormone glands. Nerves are like phone lines for carrying their messages to all the various, specific parts of the body, but hormones, such as adrenaline, deliver their announcements through a loudspeaker that goes out to the whole body at once, commanding the systems throughout the entire organism through chemical messages that circulate in the blood, regulating

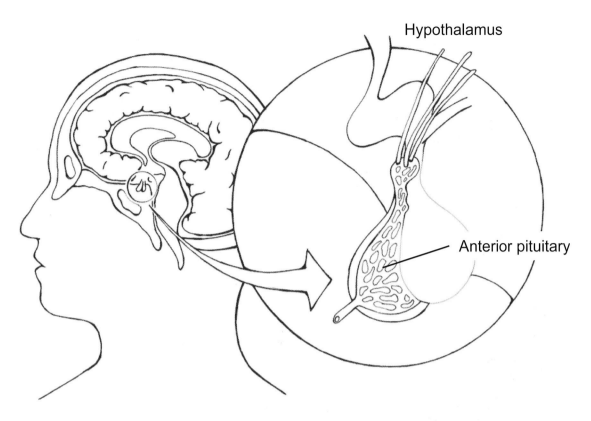

Figure 4.5

The anterior pituitary is the master endocrine gland. The hypothalamus monitors hormone levels and sends chemical signals to the anterior pituitary which responds by releasing tiny quantities of short-acting stimulating hormones that travel in the bloodstream and trigger specific hormone glands to be more active. For example, if the thyroid is slow, the hypothalamus detects less circulating thyroid hormone and releases signals that in turn drive the anterior pituitary to release thyroid-stimulating hormone into the bloodstream to stimulate the thyroid to be more active.

cell behavior and tissue functions. As the body plan is the primordial template, so hormones are the universal signaling system shared by all animals, both ancient and modern.

When wind and waves disturb the position of a small boat in turbulent waters, it pitches and rolls, working always to right itself, tipping one way only to rock back again, to settle in the center. In the same way, the body constantly monitors a great many variables and interacting factors, then continually compensates and adjusts to maintain or restore balance. Just as bilge pumps can empty

ballast from the hull of a ship that is taking on too much water, so kidneys monitor water balance in the blood and either discharge unwanted water or conserve and retain it as necessary. Sensors in the brain simultaneously regulate thirst, triggering dry mouth and conservation strategies for increasing fluid absorption from the gut when needed, while also driving water-seeking behaviors.

Beyond simply regulating fluid levels, the body also controls acid-base balance, oxygen and carbon dioxide tension, blood sugar, electrolyte balance, blood pressure, and a host of other variables – always within strict value ranges, and all as a means to maintaining an even-keel state in the body, to provide a stable internal environment for cells, tissues, and organs. This act of balancing is so consistent across multiple systems that it is a primary principle of symmetry-seeking, the necessity of balance is nowhere more crucial than with hormones. Receptors provide feedback to the brain centers that in turn moderate the release of specific hormones, signaling the body to raise or lower levels by adjusting its activities, always re-centering towards optimal values for cellular activity, organ function, and survival. And as *Homo sapiens* has adapted to life in diverse climatic conditions, the magnificent human body has evolved remarkably sophisticated hormonal mechanisms to deal with the ancient threats of hunger and thirst, heat and chill, but it is ill equipped and mismatched to deal with the modern challenges of dietary excess.

Chemical keys and cell-surface locks

Each hormone has a specific shape, like a key that will fit a particular lock, and the cell surfaces that receive the keys have matching binding sites, called hormone receptors. When a circulating hormone

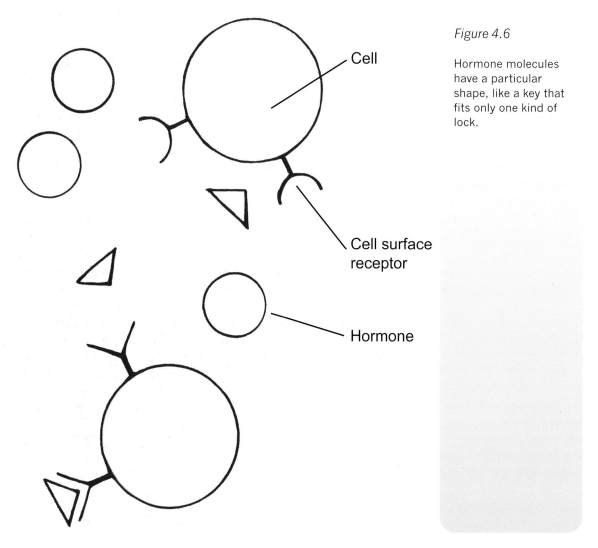

Figure 4.6

Hormone molecules have a particular shape, like a key that fits only one kind of lock.

Cell

Cell surface receptor

Hormone

binds to the receptor site on the outer surface of a target cell, it then triggers a cascade of events within the cell and changes cellular metabolism, driving specific activity or behavior.

Old pregnancy tests demonstrated the universality of these hormones across species. Today we have simple kits to diagnose pregnancy at home – a color change on a strip, triggered by hormone molecules in a drop of urine. Fifty years ago, women had no such kits, and instead doctors used the rabbit test: a drop of urine injected into a female rabbit would provoke changes in the ovaries, visible when the animal was dissected. Frogs worked too: inject a frog in the same way, and if the test was positive it would lay eggs – a better test in any case, because the same frog could stay alive, to be used again and again. This convergent response among three different creatures is an important clue. If a rabbit or frog will react to a tiny amount of hormone excreted in a newly pregnant human's urine, clearly hormones must be very potent chemical signals. Humans and frogs are vastly different; only in fairy tales can you kiss a frog and turn it into a prince. Yet hormone signals are able to talk across these widely divergent species. All species share these ancient keys, since how else could a human key fit a rabbit's lock or a frog's, to tell if you are pregnant? Hormones speak not only deafeningly loudly, but in a shared tongue, putting each species in free intercourse with the others.

Perhaps, you may be thinking, this particular hormone – the one that proclaims conception – is peculiar and exceptional. It signals a sentinel event, after all, directing an urgent change of course: that instead of shedding with menstrual bleeding, the lining of the uterus must be retained and made receptive to the fertilized egg. This hormone also signals the ovaries, thyroid, and breasts to align in concert with the new assignment and alerts the brain and appetite centers to prepare for feeding the fetal growth now anticipated. But, incredible as it may sound, other hormones are just as sweeping in their effects. They speak their universal language of commands and direct all aspects of metabolism,

from appetite and eating to mood and behavior, as well as sexual identity, bonding, pregnancy, childbirth, and lactation – not only in humans, but in all mammals.

Chemical symmetries

The first use of the word *hormone* in the medical literature can be traced to a London professor of physiology, Ernest Starling, who coined the term in 1905 to reflect his observations about the powerful nature of these chemical controls in the body – etymologically, it means 'to excite, arouse, or set in motion'.[3] The first hormone treatments were simple, long before they had a name: in ancient times, organs such as the adrenal glands or testicles were harvested from animals and fed to patients. Today, so-called 'bio-identical' therapies still offer desiccated animal glands, but now they can be marketed in tidy pill form, such as bio-identical thyroid hormone, prepared from the thyroid glands of slaughtered hogs.

Another hormone with a history of medical use is human growth hormone, prescribed for children experiencing severe growth retardation. By 1956, an extract could be derived from pituitary glands harvested from human cadavers, and this practice continued in the US for 30 years. The National Pituitary Agency coordinated the collection of pituitary tissue, as well as the creation and distribution of this rare preparation, but 1985 saw reports of four cases of CJD encephalitis (Creuzfeldt-Jacob disease, the human equivalent of 'mad cow' disease) with evidence that injections of growth hormone received when they were children had transmitted the infection to the four adult patients with CJD. Doctors in the US abandoned use of human pituitary extract and now prescribe synthetic growth hormone.[4]

Insulin is also a hormone, well known for its role in carbohydrate and fat metabolism and for regulating blood sugar levels. The insulin molecule consists of a string of only 51 amino acids, shaped like a trefoil, a clover leaf, and produced by the islet cells of the pancreas.

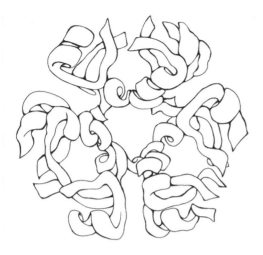

Figure 4.7

The insulin molecule has a shape like a clover leaf.

Like other hormones, insulin shows a structure that varies little between species, so both hog and cow insulin can be used in treating diabetes in humans. Hog insulin differs from human by only one amino acid substitution; bovine (from cows), by only three. And when the pancreas produces little or no insulin, diabetic patients depend on injected doses of insulin for survival. The history of medicine is rich with such clinically active, naturally occurring molecules, found in diverse species and fed back to the human patient. And, as this book will explore, hormone products like these are ten-a-penny in today's world and can come not only from animals and fish but also from certain kinds of plants.

The symmetry here – the repetition of structure and purpose – is quite striking: human hormones conversing in reciprocal dialogue with the hormones of rabbits and frogs, and horses whispering back to us through branded hormone treatments such as Premarin®, for symptoms of menopause. These ancient messenger molecules pass to and from humans, circulating master keys targeted to open or to close cell-surface locks, turning switches on or off, triggering obligatory cascades of programmed cellular activity. Hormones may seem subtle or

gentle, but make no mistake about their power. There is no more potent command than hormones for redirecting cellular activity, locking closed a gate to an existing pathway and throwing open the gate to a new path.

Because they are the most powerful biological modifiers, hormone preparations have a very important place in medical care, and usually doctors prescribe them to address specific endocrine deficiencies, such as hypothyroidism or growth retardation. Hormone preparations can treat a host of conditions, from fatigue to skin problems. Steroids are hormones, often used to counteract severe inflammation, and anti-diuretic hormone is sometimes prescribed to treat cases of excessive urine production. And in addition to these genuine medical indications, hormones can also act as performance-enhancing agents, such as those routinely making the news in various doping scandals in sports. Erythropoietin (EPO, or hemopoietin) for example, is the hormone that drives production of red cells by the bone marrow, and it is effective in treating certain types of anemia, but it has also gained notoriety as a favored illicit shortcut to increased stamina and enhanced athletic performance, as in the highly publicized cases from elite cycling.

The hormones within our own bodies offer only tiny incremental fluctuations, quite different from those introduced from the outside. The tiny, pulsed releases of our intrinsic hormones help to keep our natural biorhythms stable, regulating their physiological rise and fall within our carefully balanced, even-keel parameters. But the hormones we take in from outside – whether as part of prescribed medical treatments or unknowingly, as part of our diet or from other exposures – can have disturbing and disruptive impacts on the body. When these exogenous hormones or hormone-like substances are included in our diet, they can disturb our normal balance, creating disease and dysfunction; these are the substances known as endocrine disruptors. And a careful look at industrial US agricultural practice reveals that long-acting synthetic hormones are being used widely,

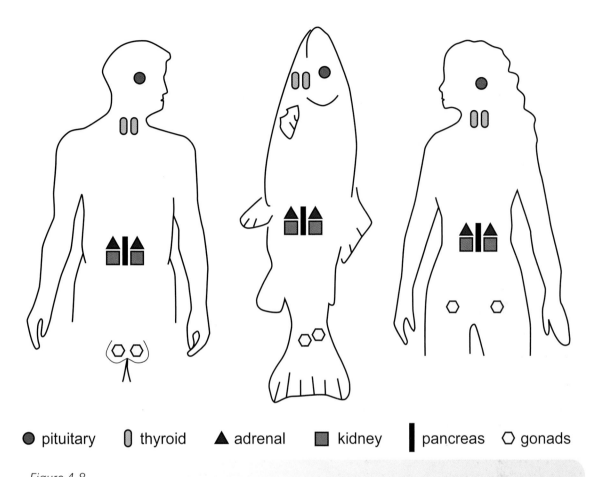

● pituitary   🮮 thyroid   ▲ adrenal   ■ kidney   ❘ pancreas   ⬡ gonads

Figure 4.8

The major endocrine organs.

particularly in the dairy and cattle industries, and that this deliberate use of hormones in agriculture presents problems, not just for a few at-risk humans but on a grand scale of consequences.

To understand how a circulating hormone, natural or synthetic, works to control cellular activity, think of each cell like a tiny jukebox. The outer surface holds the control buttons and the insides contain the records that can be played. The juke box's repertoire of song choices depends on the size of its record collection. To make it play, you press the buttons, one letter and one number, and each pairing corresponds to only that one chosen tune by one selected artist. At any moment, the jukebox might be silent, but, if triggered by having its buttons pushed, it will play one of its tunes. And even though the mechanism for making the jukebox play is always the same – drop in a coin and push the buttons – the music can vary widely depending on which buttons are pressed. Each cell in a human body is its own jukebox: the buttons are the receptors in its surface membrane, and the records are genes, ready to be turned on according to command. Hormones push the buttons. Whether our native hormones or synthetic ones that have been genetically engineered, they trigger the receptors to call the tune, activating the inner workings to demand that the cells act as directed.

Unlike our native hormones, which are very short-acting and released only strictly as needed, and then just in tiny concentrations directly into the circulating bloodstream, synthetic hormones are much longer acting, by design. In medical practice, varying preparations of synthetic hormones can be taken by mouth, inhaled through the nose, inserted as suppositories in the vagina, or rubbed directly onto the skin. They can also be given to livestock or humans by injection or by subcutaneous pellets. Animal feedstock – those raised to be food for humans – often receive hormones in massive concentrations. With so many synthetic hormones therefore entering our bodies now, both directly and indirectly, it is time to consider how they are made and why the new methods of production have made them so cheap to create and wildly overabundant.

Making enzymes

All living animals, even lowly creatures such as single-cell bacteria, have a genetic code which directs

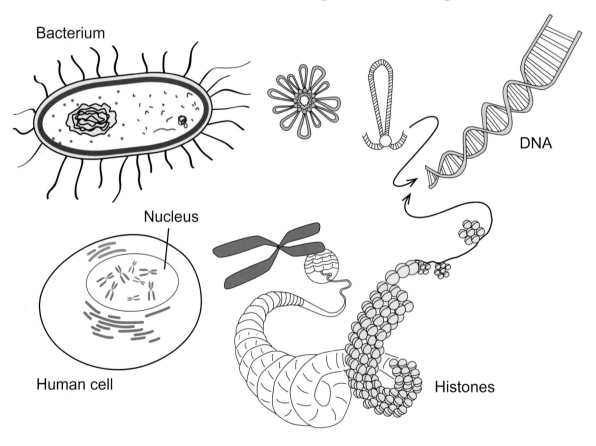

Figure 4.9

DNA in bacteria and mammals. DNA is 'free' in the cytoplasm of bacterial cells, like a single long length of string, looped and coiled on itself. In human cells, 46 strings of DNA are also looped and coiled, but grouped and stored within the nucleus of human cells as 23 pairs chromosomes. The DNA itself shares the same double helix structure, so gene segments can be cut out of human cells and implanted in bacteria, to be passed on to the next generation and the next with cell division.

and builds their structure, and bacteria play a critical role in the commercial production of hormones and pharmaceuticals, so they deserve attention as our key example of how DNA works. All DNA molecules share the same chemical structure, and the genetic code (genome) controls most aspects of metabolism and cell division, selecting what proteins or digestive enzymes to make and when to make them. Bacteria live all over the world and all over our bodies, our clothing, and our cups and glasses, feeding on simple sugars such as glucose. Amazingly, though they are the simplest of single-cell organisms, bacteria's DNA has the same chemical structure as our own human DNA.

Plentiful in the colon and in feces, most bacteria are not only harmless but positively necessary, helping us digest some of our food particles and making vitamin K and other essential nutrients available for human absorption. In illness, however, bacteria can overgrow in unwanted places, such as the vagina or the urinary tract, and cause infections of the bladder or kidneys or even bloodstream infections – septic shock.

German physician Theodor Escherich was the first to isolate and study bacteria in 1885, and the commonest kind was named after him: *E. coli*, for '*Escherichia*' and 'from the colon'. Not a single organism type but diverse families of many similar varieties, *E. coli* have been studied thoroughly, and much is therefore known about their inner workings and genetic makeup. The complete bacterial DNA sequence is a single-loop, double-helix molecule, which contains about 5,000 genes. The great majority of these genome sequences are variable, so only about a quarter of the genes are shared by all *E. coli* species, and only some genes are active at any one time, while most are switched off and inactive.

E. coli in a dilute solution of glucose, for example, do not make any enzymes to digest lactose, the 'milk sugar', so those genes are usually inactive. But deprive those same *E. coli* of glucose and place them in milk, and within minutes the bacteria will start to make lactase, the enzyme needed to digest milk sugar.

How do the bacteria know what to do, and how do they manage to adapt to new conditions so quickly? We mentioned that most genes are inactive and only some are 'open' for use. Most are covered over by an inhibitor, and when lactose is present, the milk sugar binds to the specific inhibitor and unmasks only the segment of genes that code for the enzyme lactase, allowing metabolism to adjust accordingly and proceed. This slick mechanism results not from a thought process but from the predictable activation of a genetic tool, DNA, turned on by a signal – in this case, the presence of lactose, to make the needed enzyme for digestion of milk sugar.

Though their tools are modest – simple on/off switch devices – single-cell bacteria are truly astounding in their ability to repeat this switching function across a range of purposes, as they regulate their metabolism, bacterial growth, division, and multiplication. Their simplicity also imparts startling abilities to adapt and thrive across changing conditions. The scientific knowledge of bacteria's inner workings has thus invited human ingenuity to manufacture all kinds of molecules, including synthetic hormones – active agents that can bind to receptor sites and trigger specific cascades of cellular activity.

Not all cells are equally susceptible to the effects of hormones. In an embryo, all cells are 'stem cells', with all the receptors available in every cell: all the buttons on each juke box are still in working order and every tune can be played. These embryonic stem cells haven't limited themselves with specialization yet, so they can be directed to do anything – make any protein or enzyme or differentiate and develop into any kind of specialized cell. The complete set of surface receptors on this undifferentiated cell, with their accompanying internal structures, provides the omnipotent versatility unique to stem cells. They thus attract great interest, because they can be used in all kinds of ways, to help replace or regrow any type of specialized cell in the body, such as nerve cells that have been lost because of injury or disease.

Later, as cells grow and journey in the embryo, they differentiate and diverge, with some becoming

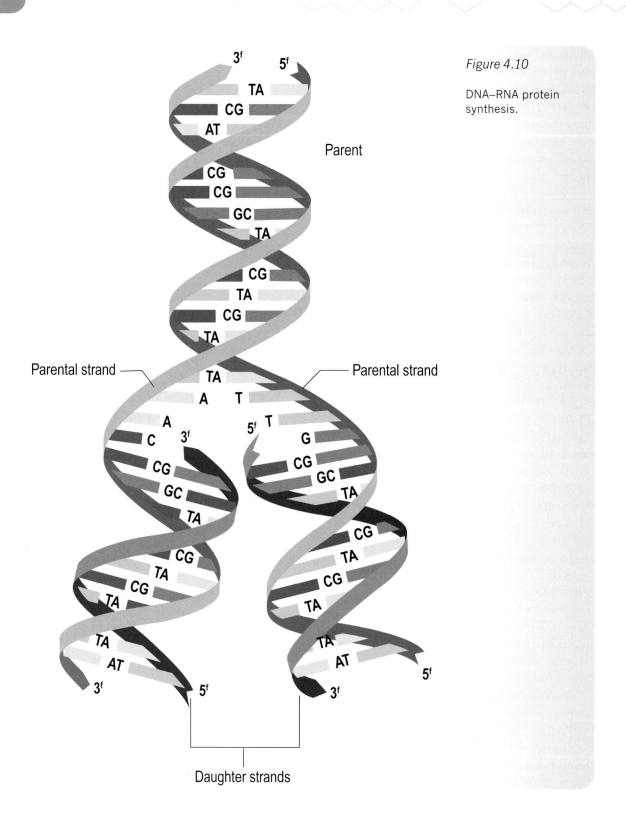

Figure 4.10

DNA–RNA protein synthesis.

bone, others skin, others inner organs. Each specializes to fulfill a given task. This process of growing new organs and novel cellular structures may seem like a series of additions, as cells acquire unique capabilities for adapting to new roles and matching needed functions, but differentiation is actually a process of loss. On each cell, some of its buttons break or stick, leaving the jukebox – the cell – less and less versatile, until only one or two songs are playable and all others are silent. When this process of loss is complete, the cell is finally specialized to its function in the body, ready to play its song. (This arc of cell development also parallels the training of a doctor, with the new medical student like a stem cell, learning about the whole body, while senior specialists become less versatile, fixed in one role.)

Making synthetic hormones

In the course of using E. coli to study and experiment with cell functions, scientists have developed genetic techniques to edit the DNA sequence. They can cut and splice to remove or add segments, like editing a reel of DNA film, thereby engineering new instructions to redirect the bacteria to new tasks. Or, to return to the idea of each cell or bacterium as a jukebox, genetic engineering can effectively change the cell's function, replacing a record or adding a new one, and when the right buttons are pressed, they trigger the E. coli bacteria to make that new molecule.

Once confined to experimental science in a research laboratory, this bacterial gene splicing has now transformed the pharmaceutical industry. Where E. coli were once simply like lab rats – an easy way to study the functions of genes and cells, due to the bacteria's versatility and easy availability – they have now become busy factory workers, performing the work of labs by cranking out replicas of themselves, in their new, edited forms, and producing limitless quantities of synthetic hormone or other kinds of molecules and drugs.

Each jukebox creates lots of tiny new jukeboxes just like itself. So, like yeast is to brewers or winemakers, E. coli is to the pharmaceutical industry. Great vats of bacterial slurry ferment the brew that pumps out synthetic hormone molecules; the bacteria are the real engines in genetic engineering.

Ordinary E. coli, the garden-variety bacteria that are common in the wild, can always be found slopping about in the human gut, but in the pharmaceutical industry they have been transformed with added segments of recombinant DNA and domesticated to live their life in an incubation tank. These new types have been genetically stirred to spit out insulin, growth hormone, or whatever biologically active molecules the inserted DNA sequence directs. In the past, hormone preparations were difficult to acquire and used sparingly, always scarce and very expensive, even compared to other high-priced medical therapies. Like a rare brandy, they always commanded top dollar in the medical marketplace. Today, the pharmaceutical industry still continues to benefit from this champagne pricing even though the bacterial method of hormone production is essentially no different from churning out barrels of beer.

Yeast was the first great giant of biotechnology. Ever present on skin, feeding on simple sugars, and responsible for the whitish bloom seen on the surface of ripe juicy fruits such as red grapes or blueberries, the yeast family of organisms has been a natural partner for humans since the beginnings of civilization, having serendipitously fallen into service for brewing beer, making wine, and leavening bread, all without charge. In stark contrast, genetically modified E. coli are newcomers born of scientific research and human ingenuity, with a story bound up in intellectual property protection, commercial avarice, pharmaceutical exploitation, and bitter competition between giant corporate rivals.

The first step on the road to genetically modified organisms was the demonstration of a technique for adding genes by splicing, achieved in 1971 by biochemist Paul Berg and Herbert Boyer of UCSF. Stan Cohen of Stanford University then improved the technique, not only adding new genes to the DNA loop of *E. coli* but also showing that the newly imported genes would be passed on to daughter cells during cell division.

In 1980, the courts strengthened intellectual property rights regarding genetic engineering, with a ruling that genetically modified microorganisms could be patented (Diamond v. Chakrabarty), and the first patent for genetically modified bacteria was awarded to one that could digest crude oil and help clean up oil spills. Bacteria remain champions today, with genetically modified organisms now so prolific that they must be color-coded using biotechnology conventions that divide commercial bacterial species into distinct groups for regulatory purposes: white for industry, blue for marine applications, green for agriculture, and red for medical uses. Biotechnology has reshaped chemical and pharmaceutical companies, as bacteria have been recruited to produce all kinds of complex molecules, including drugs and biologically active hormones – in almost unlimited quantities and at much less cost than ever before.

Experience with DES

After World War Two, science had not yet mastered genetic engineering, but it did have chemical methods of making synthetic hormones, and one of the first products was an artificial form of the female hormone estrogen. Called diethylstilbestrol (DES), it was approved for veterinary use in 1954, and the US cattle industry started using it to promote beef production. With minimal, inadequate testing for its effectiveness or safety, DES also entered clinical practice for humans, promoted for use in pregnancy as a drug to help prevent miscarriage. It was known that pregnant women with low estrogen levels were more likely to suffer a miscarriage, so doctors believed that giving estrogen might help to reduce the risk, not realizing that low estrogen was typically the result of a failing pregnancy rather than the cause. Not only did this synthetic hormone not work for its supposed purpose, but DES provoked devastating consequences for many women who used it, as well as for their sons and daughters years later. Exposed to DES during fetal life, many of them

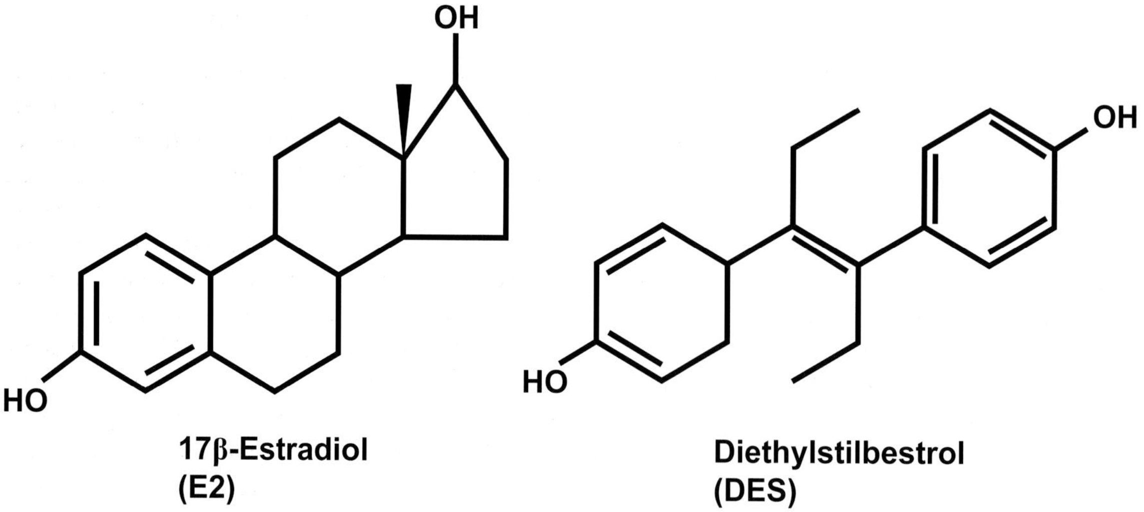

17β-Estradiol (E2)

Diethylstilbestrol (DES)

Figure 4.11

Comparison of the estrogen molecule (17β-Estradiol) and DES (diethylstilbestrol).

suffered increased rates of cancer and infertility as adults, a tragic irony for a drug marketed as a means of protecting pregnancies.[5]

The unintended downstream consequences of DES exposure in humans is an example of endocrine disruption and a spectrum of effects that spans not only congenital anomalies and both male and female reproductive disorders but increased risk of cervical, vaginal, and breast cancers. The impact of DES has now been studied thoroughly by many investigators in experiments with animals, and the results have confirmed the same pattern of increased reproductive cancer risks whether examined in mouse, rat, hamster, or monkey models. The Endocrine Society, the professional body of physicians specializing in the treatment of endocrine disorders, has published a Scientific Statement in which they propose a strong precaution: that to minimize risk, we must minimize any and all exposures to endocrine-disrupting chemicals.[6]

Hormones in agribusiness

Intimate familiarity with the power of hormones enlightens endocrinology specialists about the possible dangers from endocrine disruptors, but industrial farmers and pharmaceutical corporations approach the question from a different direction. In the 1970s, the US Food and Drug Administration (FDA) added to the list of approved hormones for use in agriculture, including two female hormones – estradiol and progesterone – as well as three synthetic analogues. These synthetic hormone drugs won this approval for agricultural uses despite vigorous opposition from the medical and scientific communities. And then in 1993, synthetic growth hormone for cattle, also called 'recombinant bovine growth hormone' (rBGH), was added to the list, promising to boost milk production. It was renamed bovine somatostatin (rBST), which disguises the nature of the drug and reduces resistance to its widespread use, since anyone can recognize what 'growth hormone' means and be concerned, whereas 'somatostatin' is opaque to non-scientists.

By 2008, one-third of all American dairy cows were being injected with recombinant growth hormone and most veal calves were pumped up with growth hormones as well.

The FDA's review process considered several studies and concluded that approval for use of rBST growth hormone was unlikely to lead to significant impact on land use, water quality, non-target species, or manure production. They also considered but dismissed the likely impacts on greenhouse gas emissions. They did not explore the potential impact on human health, and the FDA did not adequately test the milk product of rBST-treated animals as part of their approval process nor consider the possible effects on our human food chain (a subject we consider further in later chapters). The report concluded 'that the approval of NADA 140-872 is not expected to have a significant effect on the quality of the human environment', and rBST entered our food chain, unbeknownst to most Americans. Applications for agricultural use in other countries led to heated controversy; some like Canada approved its use, but others like Europe adamantly denied approval.

Hormones and the food chain

The food products that come from dairy cows treated with these controversial hormones – milk, butter, cheese, yogurt, ice cream – are only the most obvious means of passing along the synthetic hormones from the creamery into our human diet, but symmetry-seeking asks us to take a still wider view. For the agricultural industry, it may be convenient to cordon off dairy cows into a separate category from beef cattle. But for a person considering what to include in their diet, as they make their way through the supermarket, these lines are not simply blurry but a distraction from the reality that all cattle share a common path from the embryo to the slaughter house and, finally, to the kitchen and the dinner table. Dairy cattle are all female, lactating after calving and productive for a time, until natural milk production subsides. Dairy farmers take this slow-down as a cue to inject synthetic hormones to boost their milk yield, squeezing a bit more juice out of the failing dairy cow for a time. But as production drifts again and she fails to respond to even more hormones, Bessie goes not into retirement but off to the slaughterhouse with the beef cattle. The old milker's carcass – laced with freshly administered doses of long-acting synthetic growth hormone – is processed and passed on into the food chain, albeit not as prime steak for the best restaurants, but as cheap burger meat, pink slime for school lunches, and super-sized fast-food specials.

Scots enjoy haggis, the Welsh faggots, and the Polish sausage – humble 'cheap meats' using the innards rather than muscle mass – but Americans prefer to eat 'high off the hog', meaning only the best cuts of meat: shoulder, loin, and rump. Don't imagine, though, that the cheap meats don't still find their way into today's American diet, through the great many processed meat products such as pre-sliced lunch meats, burgers, nuggets, and hot dogs. Industrial processes often employ impressive and even laudable methods to maximize production and minimize waste, and modern industrial methods, applied to food, mean everything that can possibly be harvested from the carcass and presented as edible will enter the human food chain for consumption one way or another – every pig part except its squeal. In Europe, 'SPAM', the standard wartime ration, was an acronym for 'sterilized processed animal meat'; every can a new adventure, it might contain mutton, horse, or whale. Today's America may have little use for canned war rations, but slaughterhouse leftovers are still a staple of the national diet, with unmentionable animal innards, processed and reprocessed, colored and dressed up to be disguised as bite-sized delights, with no telling what you are eating.

And beyond what the slaughterhouse can make presentable as human food, the industrial mindset that seeks zero waste looks to make further use of the butchered livestock; animal processors strip muscle groups from the carcasses and transform the remains into commercial products such as tallow, bone meal, and animal offal, for use in dog food, cat food, fish meal, and even commercial chicken feed. So the hormones prohibited for uses in poultry farming but allowed for fattening cattle and hogs still find their way into other human foods. The animal meal used as a food additive for industrial chicken farms, for instance, creates a route for synthetic growth hormones to contaminate chicken products. Industrial farming effectively redistributes excessive synthetic hormone residues from cattle and hogs not only through cheap processed meat products but also through chicken and some farmed fish, such as the voracious tilapia, thereby building multiple routes for these hormones to enter the human food supply.

Joining the dots, across species

Adolescent girls grow faster than boys because female hormones drive growth, and we eat a great deal of food filled with female hormones. All dairy and almost every piece of meat consumed today is derived from the female. Whether you are eating beef or pork, the source animal is almost always female or a castrated male (so, without male hormones). And you are what you eat. Just as estrogens drive growth of secondary sexual characteristics in females, if males receive enough female hormones or hormone-like chemicals, their breasts will also grow and develop, like female breasts, because the toolkit of master genes for cell growth is the same in both men and women.

Children are dramatically more sensitive to the effects of hormones than adults, their young cells like the shiny new jukebox, with its buttons untouched but fully equipped and ready to play. That young farm animals show this same exquisite sensitivity, industrial farmers manipulate to their advantage: in the same two years it takes a human infant to toddle, corporate producers can pump up the weight of a beef calf from 80 pounds to a market-ready size of more than 1,000, if they spur the process with liberal use of synthetic estrogen hormones. Administered deliberately to distort biological conditions and beef up weight gain, with total disregard for downstream and distant hormonal fallout, this intentional short-term benefit for agribusiness brings unintentional negative consequences, including the not-distant reality that human children, as super-sensitive as young calves, will shortly be ingesting the hormones downstream.

Remembering that even the tiny concentration of hormone in a pregnant woman's urine triggers the frog or rabbit to ovulate, symmetry-seeking asks us to broaden our perspective, to consider possible similar impacts from other hormones, across species and in other contexts. If we know not only that hormones in human urine create change in other species, but also that horses' urine is a common source for hormones given to women for menopause symptoms, what might we now learn about human susceptibility to long-acting synthetic hormones in the human food chain? The ancient masters of growth, the Hox genes, are still directing our growth and the development of all mammals, so we cannot ignore that primordial hormones at work in one kind of animal will have some activity in all mammals, including humans. How high, then, might be the human costs we are paying for cheap meat?

Speeding the pace of growth and weight gain accomplishes its intended benefits for the industrial farmer, who gets more yield for less feed, a shorter timeline to an earlier sale, and an early start on feeding the next round of young calves for the next cycle in the rush to market. The farmer may be unconcerned that long-acting synthetic hormones are actively excreted in the urine and that, beyond the farm, they might be passed on as downstream runoff, spilling over to fatten and feminize children, disturb sexual function and fertility, or hasten the pace of tumor growth.

Other nations have considered the evidence and banned the use of these growth-promoting, long-acting synthetic hormones and the foods that contain them. Many countries have also banned US meat and dairy products, which cannot be imported into Europe because of concerns for human health. But in spite of all that is known, synthetic and biologically modified hormones are still widely used in US agriculture, and meat and dairy products are marketed without adequate labeling, consumed by families who have not been informed of the potential risks. Meanwhile, the US Department of Agriculture promotes hormone use, and the FDA joins with the cattle and dairy industries who profit from this ill-thought-out expediency, continuing to assure consumers that long-acting synthetic hormones are safe, despite the compelling evidence to the contrary. In light of such ill-considered public policy, as we will explore in the later chapters of this book, a wider view of developmental symmetry between humans and farm species carries clear implications not only for responsible agribusiness but also for personal daily food choices.

References

1. Fink T, Mao Y (2000). The 85 Ways to Tie a Tie: The Science and Aesthetics of Tie Knots. New York: Broadway Books.
2. Bejan A, Peder Zane J (2012). Design in Nature: How the Constructal Law Governs Evolution in Biology, Physics, Technology, and Social Organization. New York: Doubleday.
3. Henderson J (2005). Ernest Starling and 'Hormones': an historical commentary. *Journal of Endocrinology*, 184(1): 5–10.
4. Frasier SD (1997). The not-so-good old days: working with pituitary growth hormone in North America, 1956 to 1985. *The Journal of Pediatrics*, 131(1): S1–4.
5. McLachlan JA (2006). Commentary: prenatal exposure to diethylstilbestrol (DES): a continuing story. *International Journal of Epidemiology*, 35(4): 868–870.
6. Diamanti-Kandarakis E, Bourguignon J-P, Giudice LC, et al. (2009). Endocrine-Disrupting Chemicals: An Endocrine Society Scientific Statement. *Endocrine Reviews*, 30(4): 293–342.

Symmetry reveals shared targets and drug actions

Hormones represent only one class of biological modifiers that were once beyond the reach of human interference and now accessible to biopharmaceutical manufacture. Penicillin, the first antibiotic, was discovered by Alexander Fleming in 1928 and developed during World War Two, and it has not only cured many an incidental infection but saved countless lives. As with hormones, however, penicillin's widespread overuse has led to a host of complications.

By the end of the 1940s, antibiotics were still precious, few, and rarely prescribed. But in the decades since, use of antibiotics has become copious, unrestrained, and ill-considered, not only in humans but in animals, as US agribusiness has become the Wild West of reckless antibiotic use. Many and dangerous new strains of resilient bacteria have appeared, such as MRSA (methicillin-resistant *Staphylococcus aureus*), a bacterium armed with penicillinase enzyme and thus resistant to penicillin's action. And once a body acquires the resistant bacteria, they flourish and blossom to become more prevalent. The clear cause and promoter of antibiotic resistance is the overuse of antibiotics.

Fleming himself had foreseen the potential dangers of antibiotic resistance as early as 1945, before the end of World War Two, during which thousands of the wounded were saved from infections that would have been fatal without antibiotics. As he explained the risks of overuse to *The New York Times*, 'microbes are educated to resist penicillin and a host of penicillin-fast organisms is bred out... In such cases the thoughtless person playing with

penicillin is morally responsible for the death of the man who finally succumbs to infection with the penicillin-resistant organism. I hope this evil can be averted'.[1] Unfortunately, however, such deaths from resistant infections are increasingly common in American hospitals now, in the 21st century.

The problem of antibiotic resistance is a lifethreatening reality for patients and a daily challenge for hospital doctors, who have little control over their colleagues' prescribing practices and absolutely no say in the commercial decision-making of agribusiness. We are all in this together. Antimicrobial use in agriculture has profound impacts on human health and survival, but farming and doctoring are leagues apart, and any linkage between them has been lost. Their separate theaters provide totally different viewpoints and perspectives, and what is obvious to experts in one industry is invisible to experts in another.

Tracking antibiotic resistance

Smaller countries are far ahead of the US in collecting and benefiting from systematic record-keeping on antibiotic consumption. Europe has both institutional memory and tangible reminders of the plagues that once devastated the city populations of London, Paris, Rome, and Venice, so it is little wonder that European countries take the threat of widespread infectious disease more seriously than does the US.

Denmark has a strong agricultural tradition and was among the first to create an antimicrobial resistance-monitoring program, which continuously tracks antibiotic use and resistance in humans, animals, and commercial food products. The record-keeping includes all relevant parties, so that data from doctors' practices and from farmers come together in the same arena, gathering statistics on medical prescribing for humans and antimicrobial consumption in agriculture, presented side by side, where researchers can then recognize patterns and links.

In a multi-nation effort, the European Surveillance of Antimicrobial Consumption Network collects data from 34 countries about human and animal use of antibiotics, and its companion the European Antimicrobial Resistance Surveillance Network collects, studies, and reports on changing patterns of bacterial resistance. In the US we monitor other potentially life-threatening phenomena, such as weather conditions, employing that information to predict tomorrow's storms with increasing accuracy, because it makes sense that early warnings, when heeded, prepare us to better meet their threat. But we have no adequate system for monitoring antibiotic consumption and controlling bacterial resistance, and until we do, there will be no way to predict or contain the looming epidemic of drug resistance in the US – and no tools available to limit loss of life and devastation.

Urinary tract infections are the most common bacterial infections, with about 150 million worldwide every year. In the US, this infection prompts about 8 million physician visits each year, occasioning significant health care costs and morbidity. Our old friend *E. coli* is the commonest pathogen to cause infections in the urinary tract, since it is already present in the gut of both humans and animals. When it is released into the environment it can find its way back to the body, to places where it is unwelcome or dangerous. In urinary infection, *E. coli* often passes from a person's bowel to their bladder, via the neighboring orifices of anus and urethra. And in food poisoning, the bacteria in trace amounts of fecal matter can pass from host to host in food, or from the hand to the mouth. (The washroom signs in

US restaurants directing food workers to wash their hands with soap and water after using the toilet are an effort to limit this fecal-to-food transfer.)

A veritable reservoir for antibiotic-resistance genes, *E. coli* accounts for up to 90% of women's uncomplicated lower urinary tract infections, and typical medical treatment for a non-pregnant healthy woman is a single dose or at most a 3-day course of a first-line antibiotic. If the bacteria are resistant, however, this basic treatment does not work. In just 10 years during the late 20th century (between 1989 and 1999), the rate of resistance more than doubled, from 8% to 18%. Urinary tract infections can be treated effectively without antibiotics, by drinking lots of water to dilute the germs and urinating more often to displace them. Taking vitamin C can help too, by acidifying the urine and slowing the pace of bacterial growth. And paying attention to bowel function, especially by resolving constipation, hastens the resolution of lower urinary tract symptoms. Cranberry juice is a traditional therapy that works well for some patients, so the active ingredient in cranberries has been isolated and manufactured, available without prescription.

Antibiotics tend to be the default treatment, though, and when first-line antibiotics fail, the physician's next step is often to double down with newer and more powerful antibiotics, such as drugs from the fluoroquinolone family (ciprofloxacin or levofloxacin). When new agents first appear, each enjoys some early successes, but bacteria are worthy adversaries. Once bacteria meet an antimicrobial, it isn't long before they have re-engineered their own defenses to resist it. And bacteria's ability to resist is accelerating. In just twenty years since their introduction, the promise of newer antibiotics has already begun to evaporate, going from very useful to almost hopeless. In Georgia, where I practice urology, resistance that was less than 1% when the fluoroquinolone drugs were first introduced has already surpassed 35%.

Clostridium difficile (C. diff) is a type of bacterium commonly found in the human gut. There it competes with its neighbors like *E. coli* for a place, in the same way that plants and weeds fight for space in the same field. When used, antibiotics such as fluoroquinolones kill off the sensitive bacteria but not germs like *C. diff*, which are much more resistant to common antibiotic therapy. The overall result is thus a disturbed balance in the ecosystem of the gut. Fluoroquinolones eliminate natural competitors, leaving the *C. diff* to bloom and blossom and fill the space. Overgrowth of *C. diff* can cause life-threatening diarrhea, and 250,000 *C. diff* infections occur in the US every year, accounting for more than 14,000 deaths and at least $1 billion dollars in additional health care costs per year. These *C. diff* infections are more common than ever before in US hospitals, and the death rate from *C. diff* climbed fourfold between 2000 and 2007.[2]

Starting with Jenner, Pasteur, and Lister less than 200 years ago, our response to deadly bacterial diseases has travelled a roller coaster from total therapeutic helplessness to triumph, but we are now looping back towards helplessness again. How did we manage to squander such an asset as fluoroquinolones in double-quick time?

The answer lies in industrial agribusiness. This antibiotic quickly became a cornerstone of effective medical therapy, prescribed for treating human illness, but because it was cheap to manufacture it was also sold on the open market to agribusiness, where it is routinely added to animal feed and chicken water. A staggering proportion of antibiotic use in the US – more than 70%, by conservative estimates – is not put to treating human disease but squandered in commercial livestock production,[3] where industrial methods and overcrowded conditions of confined animal feed operations necessitate their use. A phenomenon we'll consider within the larger context of animal farming, in Chapter 7, this reckless practice has been strongly criticized for years by the medical and scientific communities and is known to be the major driver of increased antibiotic resistance and increased death rates from infections, seen in general medical and hospital practice.

Cycle of death from bacterial infections

We had virtually defeated the infections that most threaten human life, with penicillin's vital success founded on a unique asymmetry: it targets a physical structure present in some bacterial cells but never found in human cells. By attacking only that structure and its key metabolic pathway, penicillin could poison and kill the offending organism without poisoning the human host.

Bacteria are single-cell organisms, each protected by its own kind of waterproof coating, like a protective shell that wraps around its outer boundary. In the round-shaped bacteria such as staphylococci and streptococci, penicillin interferes with the building process that creates this outer shell. The thick outer coat of the cocci includes a network of interlocking structural protein molecules (peptidoglycans) that provide rigidity and support to the waterproof coat. Throughout the bacterial lifecycle, the cell wall is constantly remodeling to accommodate cycles of cell division and multiplication, and protein molecules in the cell wall hold it together, like an outer capsule around the goo inside. Penicillin blocks their

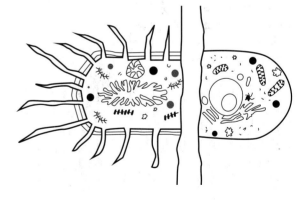

Figure 5.1

Comparison of a bacterium (left) and a human cell (right). Cellular structures are present in both, but only the bacterial cell is encased in an outer wall.

cross-linking activity and effectively disrupts new cell wall construction. Without the insulation of the outer wall to waterproof them, the bacteria are vulnerable to fluid flux, so they take in excessive water and swell up until they burst open like grains of rice.

Human cells do not have an outer coat, because they don't need one; safely enclosed within our body, already protected by our outer skin, most of our cells have no demand for a waterproof outer wall of their own. Penicillin is able to exploit this unique asymmetry between the bacterial cells and ours to attack the key structural elements critical for bacteria but not present in a patient's cells. This difference is the reason penicillin can poison and kill the bacteria without damaging the human host.

Other bacteria require different treatment strategies. *E. coli* and other gut bacteria, for instance, are rod-shaped and constructed a little differently from the round-bodied cocci: they have a thinner outer cell wall, with more fat (lipopolysaccharide) and less protein in the outer coating. Penicillin is less effective against these rod-shaped organisms, which are typically responsible for urinary tract infections or infectious diarrheal illnesses, like food poisoning, than it is against the kind of bacteria (streptococci and staphylococci) that can cause sore throats or pneumonia. So according to the specific type of

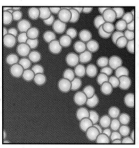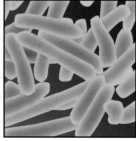

Figure 5.2

Round cocci (left), may be single or in pairs, arranged in strings or bunched together like grapes. Rod-shaped bacteria (right) are most commonly found in the gut.

bacteria and their particular structural elements, one or another antimicrobial will be more or less effective, and if the germs have already survived prior exposure to an antibiotic, they will have acquired some degree of resistance to its action. Since the discovery of penicillin, scientists have developed an array of antibiotics to counter the variety of microbial infections, but each new drug initiates again the cycle of resistance to itself.

Bacterial diseases have plenty of opportunities to practice their resistance, because these infections can be spread in any number of ways – borne by air, water, and food, especially. Common infections, such as the periodic outbreaks of salmonella or *E. coli* that we call 'food poisoning', come from tainted foods such as cold meats or undercooked chicken and burger meat. Once readily manageable with first- or second-line antibiotics, food-borne infections have become more likely to be fatal. In addition to instructions in handwashing, we teach children that 'coughs and sneezes spread diseases, catch them in your elbow', since coughing and sneezing expel fluid secretions and germs from our bodies into the air and onto our hands or other surfaces. But many bacterial organisms that cause diseases do find their way to a human host, and they carry a new level of threat, if they have already been exposed to antibiotics on the farm. It is the toughies that have won the game of survivor, by acquiring the power to resist the actions of the antibiotics that could kill them.

Unfortunately, the vast majority of metabolic pathways are shared between bacterial cells and mammal cells, just as the simple, classic chemical structure of DNA is exactly the same in both. Our healthy human cells and the organisms that cause infections are therefore too alike in the way they work for both not to respond similarly to shared chemical attack, and there lies the greatest challenge to confound the search for the next new class of antibiotic agents. How to poison the germ without also harming the human host? At a loss for answers to this question, after unleashing the power of bacterial ingenuity on the world, the impotent pharmaceutical industry cannot begin to match bacteria's vigor and versatility. Bacterial resistance to treatment is the scourge of modern hospital medicine, and currently more Americans die of antibiotic-resistant bacterial infections than of more publicized health epidemics caused by viruses, such as HIV and AIDS.

Beyond antibiotics: the clumsiness of drugs

If we step back to look beyond antibiotics to the effects of medications more generally, we face exactly the same challenge, in the quest to find drugs that sustainably, effectively do more good than harm. The symmetry between cells creates the challenge: the cells of the body are more the same than different, so any drug carried in the bloodstream – which is every drug – will affect cells throughout the body in similar ways. Every cell in our body shares the same ancestor. The fertilized egg cell was the stem cell that divided into daughters again and again. If some of our cells become malignant and grow into a cancerous tumor, the cells of the tumor will be very similar to the cells throughout the host, much more similar than different. So chemotherapy drugs designed to block metabolic pathways and kill the cancer will also risk killing your own healthy cells – the reason your hair falls out, your immune cells die, red blood cells expire, and you become sick and anemic. If the cancer is feeling the effects of chemotherapy's poison, the whole body will feel it as well.

Drugs in the body are subject to breakdown and metabolism like any other molecules. When we take medicines by mouth, the process of breakdown is the same as for food. The stomach and gut digest and absorb the chemicals in the drugs, passing them through the portal bloodstream directly to the liver, whose job it is to filter out anything harmful to the body.

Along the gut tube, the liver takes the first hit of toxic substances, as it attempts to filter them out before they reach the rest of the body. Certain substances have long been known to overload the liver's ability to do its job, such as alcohol when it is regularly consumed to excess. Such overload will cause inflammation and scarring – alcoholic hepatitis and cirrhosis. But more than half of all cases of acute liver failure in the US are caused by FDA-approved prescription drugs or over-the-counter medications; daily use of acetaminophen (paracetamol) and other medications can cause liver damage and even liver failure. More than 900 drugs and herbal products are reported to cause liver damage. Drug-induced hepatic injury is the most commonly cited reason for withdrawal of an approved drug by the FDA, and severe liver injury due to drugs is more likely to lead to liver transplantation surgery than alcoholic cirrhosis.

Just as the liver is the organ that filters the blood flow from the gut, the kidneys process the blood from the rest of the body. They filter out any toxins and waste products that remain after nutrients and other molecules have circulated through the body tissues and organs, in the bloodstream. The kidney tubules clear most medicines from the body, eliminating them in the urine, so many drugs end up concentrated in the kidney tissue and can cause injury to the kidneys. Common families of toxic drugs include antibiotics, blood pressure medicines, and chemotherapy drugs. Antiviral agents, the contrast media used in medical imaging, medications for diabetes or even indigestion, and over-the-counter anti-inflammatory drugs can cause kidney injury on their own, but they are a greater risk in combination – not to mention common household chemicals such as cleaning products and garden sprays, including insecticides and weed killers.

A drug that is directed to have only one clinical effect, such as bringing more blood flow to the penis for sex, will also bring extra blood to other body parts, especially those in corresponding, symmetrical systems. A structure linked to the penis is the face, in particular the cheeks, nose, and nasal passages – hence the unwanted facial flushing and stuffy nose that we saw from erectile dysfunction drugs, in Chapter 3. Spontaneous physiological sexual arousal is also likely to be accompa-

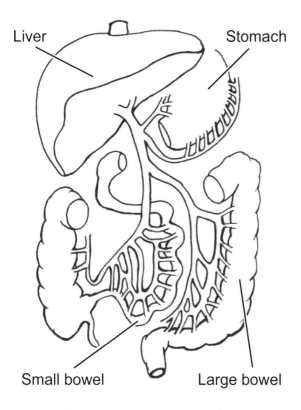

Liver

Stomach

Small bowel

Large bowel

Figure 5.3

Blood flow from the gut to the liver. Venous blood drains from the gut tube to the liver through the portal venous system. The liver acts as a filter to remove toxins absorbed from the gut. Medications taken by mouth will be absorbed and first pass through the liver before entering the venous blood system to return to the heart; that blood passes next through the filters of the lungs before returning to the heart to be pumped around the body from the left ventricle. Food and drink follow the same path so excess alcohol will lead to liver damage (alcoholic cirrhosis of the liver).

which will bring both the wanted and unwanted effects together, performing its intended task at the one end of the body and inviting unwelcome actions at the other.

Structures built the same way in the embryo share the same signatures and offer the same targets to circulating compounds, whether the chemical arrows arise within the body or from without. Thus, medications prescribed for symptoms of overactive bladder will tend to cause dry mouth and constipation, because the muscle of the bladder wall is so similar to the muscle of the gut and salivary glands. Like structures mean like responses to medication, and once in the body the active agent will travel everywhere it can go and perform its action on any and all receptive tissue targets. Usually thought of as 'side effects', these unintended consequences can be more accurately understood as simply effects, as these similar events are the direct results of medication, the intended actions in parallel but unintended organs. And the more potent the drug for its intended task, the more powerful and widespread will be the unintended hits.

To do good or to do well?

Biopharmaceuticals are typically large molecules, akin to a human's own antibodies and hormones. By activating our specific cell-surface receptors, these drug molecules initiate, accelerate, or slow certain biological processes. Because these compounds are large and acid digestion in the stomach can break them down, rendering them less effective, biopharmaceuticals are typically administered by injection, thus bypassing both the stomach and liver (whose role it is to protect us from toxicity). To produce medicinal molecules ranging from antibiotics to insulin, modern biotechnology uses genetically modified microorganisms, such as the familiar *E. coli* and yeast, but now there are also new recruits to the pharmaceutical workforce. They include genetically modified mammalian cells, including Chinese hamster ovary cells, and vegetables such as bioengineered corn.

nied by some heightened color in the face, but drug-induced erection can intensify the effect considerably, to the point of discomfort. Such mirror events reflect the dual actions of any medication,

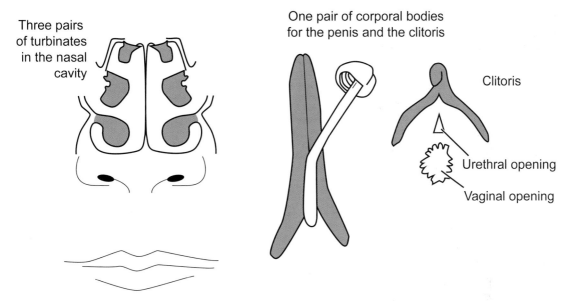

Three pairs of turbinates in the nasal cavity

One pair of corporal bodies for the penis and the clitoris

Clitoris

Urethral opening

Vaginal opening

Figure 5.4

External projections of nose and penis/clitoris. Vascular spaces flank the external openings of the respiratory and urinary tracts, as in the turbinates of the nose and the corporal bodies of the external genitalia. The vascular tissues are fused to bone on the right and left. Three pairs of vascular turbinates serve as air conditioners to warm and humidify air breathed in through the nose. One pair of corporal bodies provide the vascular cushions of erectile tissue that project forward in the midline for the penis and clitoris. Hair growth is relatively sparse on nose, penis, and clitoris and the coverings are not only erectile, but rich in highly sensitive and responsive erogenous tissue.

For biopharmaceutical companies, money may not grow on trees, but it does grow quite easily and cheaply on bushes and other plants.

To be a blockbuster moneymaker for a biopharmaceutical company, a drug will treat a condition that requires lifelong management or at least long-term treatment, not a single dose. Examples of such creations include etanercept and infliximab, two different types of drugs (a fusion protein and an antibody therapy, respectively) both used to treat Crohn's disease, rheumatoid arthritis, and other autoimmune diseases. Another success for its developers is rituximab, an antibody used to treat lymphoma, leukemia, and episodes of threatened rejection after solid organ transplantation. Biopharmaceutical companies have also developed some new molecules that improve diagnosis and guide treatment options, such as herceptin, for breast cancers that express the protein HER2.

Some of these advances are truly new landmarks in medical treatment, as are new molecules created in laboratories to treat hepatitis, cancer, or hemophilia. Modern biotechnology can also make existing medications more easily than before and relatively cheaply; in 1978, the company Genentech successfully inserted the genes for human insulin into a strain of *E. coli*, opening the door to their producing bio-identical insulin in vast quantities at relatively low cost. The other insulin products are still derived from animals and easy to harvest, procured from the pancreas glands of abattoir-slaughtered cattle or pigs.

The newer, synthetically produced human insulin actually offers little clinical advantage over the modern, purified version of porcine insulin, because pig insulin is so very similar to human, but the biopharmaceutical creation still demands, and receives, twice the price in the medical marketplace. Even though science has long since overcome the once-formidable technical obstacles to using *E. coli* in making 'human insulin', patent laws nevertheless allow biopharmaceuticals to drive the costs of medical treatments to new, unthinkable, and unsustainable heights, attracting both national and global attention and now the subject of ongoing inquiries in both US Congressional committees and the British Parliament.

There is no doubt that medicines do contribute enormously to health. For many people, the discovery and development of effective drugs has greatly improved quality of life, avoided the need for surgery, reduced time in hospital, and saved many lives. And pharmaceuticals have become such a default treatment for all manner of conditions that, in the UK, about 650 million prescriptions are written in family practice alone, for drugs costing more than 7 billion pounds sterling annually, with 80% spent on patented (brand name) products. In Britain, the pharmaceutical industry is the third most profitable economic activity after tourism and financial services. In the US, spending on prescription drugs is astronomical, with more than 4 billion prescriptions written in 2011, and still rising rapidly. Americans spent $450 billion in 2016,[4] and the total is expected to rise to as high as $610 billion by 2021. Among adults, 46% take a prescription drug, with 11.5% taking three or more every day.[5,6]

We have been lulled into thinking that the interests of pharmaceutical industry and the public are the same, but they are not. There is overlap between success for drug companies and increased health for those who take their medicines, certainly, but their goals do not reliably align. Pharmaceutical companies profit when more people use and rely on their drugs, but using and relying on pharmaceuticals is by no means a reliable path to better human health,

overall. And the risks inherent in these medicines are considerable, for any individual: adverse drug reactions, toxicity, drug interactions, ill health, and even death.

The pharmaceutical companies' vested interests are now inextricably interwoven with big-picture decisions at every level of health care, far upstream of an individual doctor's decision on whether and what to prescribe to an individual patient. 'Big Pharma' chooses which drugs to develop through clinical trials and which to promote to prescribers and patients; they even share in developing clinical practice guidelines for physicians, built around the drug companies' products. And at every step of this journey, aspects of the industry's interests diverge starkly from patients' interests. Clinical trials are often poorly designed, to meet minimum legal requirements and avoid learning about a drug's true risk profile or the limits of its effectiveness. Driven by profit interests rather than scientific objectivity, pharmaceutical companies cherry-pick data to publish about trials, displaying results in the most favorable light, selectively choosing what trials to publish, and having company representatives ghostwrite the results for scientific journals. Together, these strategies suppress information about adverse effects, so that published evidence about a new drug may fail to reflect its true benefits and risks.

Once medications are approved, their patent-holders promote them intensively, not only to physicians but also in direct-to-patient advertising.[7] Patient organizations spring up to generate media attention around their ostensible 'disease' and drive patients to demand the new drug from their doctors. Falsely framed as grassroots patient-advocacy groups representing a genuine and underserved patient population, these organizations often turn out to be created, organized, paid for, and scripted by sales and marketing departments of the pharmaceutical companies themselves, earning them the nickname 'astroturf organizations' – fake grass, with fake roots. Like other industries, the pharmaceutical business can thus become ever-more driven by its marketing department. And in health care, the larger market is not the sick but the healthy, since many overall healthy people have mild-to-moderate symptoms of ongoing conditions – many more than those with moderate-to-severe diseases. 'Wellness' rather than 'illness' then becomes the buzzword of Big Pharma, while they market their products as necessary to maintain health, not simply to treat the sick.

And when wellness fails to drive sales strongly enough, the industry is glad to characterize a range of symptoms as illness, contributing to the mistaken idea that more and more conditions are diseases and that every problem requires a medical treatment. Two drugs presented as wonder drugs for arthritis illustrate the problem well: Vioxx® and Celebrex®. These COX-2 inhibitors were marketed and vigorously promoted as wonder drugs for symptoms of arthritis, but were withdrawn after a few years in 2004, as thousands of deaths and many more cases of heart failure were linked to their use – conditions much worse than the symptoms they were marketed to treat. Investigations afterwards in both the UK and the US revealed many inadequacies in the approval process, as industry had not disclosed unfavorable trial results and had failed to follow appropriate licensing procedures or the necessary post-marketing protocols (for tracking responses to a drug once people begin to use it).

The business of pharmacies

The roots of pharmacy as a profession lie in the 17th century, when King James I established the guild of apothecaries. Early pharmacists prepared and dispensed remedies and offered medical care to their customers, and they traveled to the New World with English colonists, where they flourished and diversified. In an effort to standardize the field in the mid-19th century, Edward Parrish proposed the name 'pharmacists' and led the establishment of the American Pharmaceutical Association. For the next century, pharmacists' role was not only to make and provide medicines but to prescribe them, as well as to provide first-line medical care. Then in 1951 the Durham-Humphrey Amendment to the Federal Food, Drug and Cosmetic Act of 1938 created guidelines for prescription and over-the-counter medicines that required a physician's prescription for many medications and caused a shift in pharmacists' focus, toward dispensing and patient education.[8]

Up until World War Two, then, it was the local pharmacist who compounded most prescription medications. Remember the movie *It's a Wonderful Life*, when the young George Bailey notices a lethal error and brings it to Mr. Gower's attention? In those days, with far fewer drugs available, the pharmacist filled each prescription, making it from raw ingredients and packaging to suit an individual patient's needs. Today, 'compounding pharmacies' are specialty shops for unusual medicines, but through the first half of the 20th century, all pharmacies were, by definition, compounding pharmacies. However by 1950, after the war, mass-produced bulk products from drug and chemical companies had replaced most pharmacy-filled prescriptions. Such standardization was an advance for quality assurance, but it also opened the floodgates to pharmaceutical commerce, including innovative development and enormous profits, beyond the neighborhood pharmacist's imagination. Economically, the pharmaceutical industry became a major success story, and it has also made some remarkable advances in the struggle against illness and disease. The industry does deserve congratulations for its many genuinely positive achievements.

But, to be clear about their business model, neither the local pharmacy nor Big Pharma is invested in your health. What keeps them in business is your illness, or at least your perception that you have an illness – preferably, a chronic one. Your neighborhood Walgreens may advertise itself as sitting 'at the corner of happy and healthy', ready to help you and your doctor make you healthier, but a good doctor's advice is likely to have included behavioral interventions that have no retail value in the corner drug store. If she is attuned to the true sources of health, your physician may have prescribed not only that you take this medicine but also that you exercise regularly, start getting enough sleep, and of course eat real food, mainly fresh vegetables and fruits. If the pharmacy business model was aligned to promote health, Walgreens would promote a healthy diet and sell fresh vegetables and fruit. Instead they peddle candy, cigarettes, soda, tobacco products, and sometimes alcohol or processed produce, no different from a gas station. Kudos to CVS, which recently withdrew cigarettes and tobacco products from their shelves, but don't expect to find healthy food there either.

This obvious disconnect between pharmacies' ostensible purpose and what they actually sell reveals the paradox of a stated investment in health, with an actual, deeply vested interest in continuing illness. With so few profits to be gained from customers following a healthy diet and lifestyle, the purveyors of pharmaceuticals show no desire to lead the way.

Today's marketplace has changed dramatically from the small, independent corner drug store of the mid-20th century. With all they sell, both from behind the prescriptions counter and in their other departments, pharmacies have become big, big business. Some pharmacies offer the extra convenience of a drive-through window, like fast food restaurants, and the major supermarket chains have pharmacies within them, so as not to pass up on their share of this lucrative market. Among Fortune 500 companies, the pharmacy chains CVS and Walgreens are both in the top 20. At number 7, CVS holds an estimated net worth of more than $153 billion, and Walgreens/Boots at number 19, comes in at over $103 billion. Walmart is number 1 at $482 billion, Target number 38 at $74 billion, and RiteAid is valued at $26 billion. Retail goliath Amazon is reportedly seeking to break into the pharmaceutical market so as not to miss out in this immensely profitable sector of the US economy. We spend a great deal of money in the pharmacy, and often.

When we compare the businesses of food and of pharmaceuticals, one aspect of selling groceries and selling drugs remains notably asymmetrical: the checkout experience. Grocery items are labeled with standard bar codes, to be scanned and recorded. For the seller, bar codes are dense with information about the product – origin, supplier, quality rating, tracking, and date of harvest or manufacture. So while big food industries have fought efforts to improve food labels for consumers, to tell us what ingredients and nutrients are in our food (a problem we'll return to in Chapter 8), these companies have welcomed the efficiencies of barcodes to provide data linkage for suppliers to match production with demand, for distributers to match inventory with commerce. For customers, barcodes carry the benefit that every customer pays the same price for the same item, at the amount prominently displayed alongside the item on the shelf and its name on a receipt.

But data tracking for pharmaceuticals remains inadequate for both seller and buyer. Barcodes are still very rare in pharmacies, which makes vitally needed data about pharmaceuticals sparse, as is information about their origins, including what country they come from. And in comparison to foods, labelling requirements are utterly trivial for minerals and vitamins sold over the counter, leaving a great deal of critical information generally unknown to consumers. And pricing for drugs is an utter free-for-all, where walking up to the cash register in the pharmacy corner of the store can feel like Russian roulette. A drug's cost varies dramatically based on the purchaser, not the product. Like today's hospital charges, pharmaceutical pricing ranges widely for patients according to insurance status, networks, coverage plans, and other unexpected variables, such as availability of generics or the lack thereof, or for reasons undiscernible to the customer, such as where the drug stands in its patent cycle, for the company who owns it. Unlike grocery products, which arrive in stores with recommended retail pricing and can be compared easily to prices in online stores, drug prices are now opaque. But it is consumers – patients – who pay the price.

Collateral damage

Whether regulated to be prescribed by doctors or sold over the counter, medications are and have always been double-edged swords, able to cut both for better and for worse. And as professionals, both pharmacists and physicians have always recognized medicine's poisonous effects. They are trained to be aware that all medications are carried in the blood-stream to reach all parts of the body and can have unwanted consequences. General efforts to stand-ardize the training and quality of pharmacists can be traced back to the first guilds, and concern arose later for the safety of specific drugs – awareness that commercially produced pharmaceuticals might not be simply ineffective, but dangerous. In Britain, the first national scandal about the safety of a new drug arose in 1848, in Newcastle. During a procedure for an in-growing toenail, a young girl named Hannah Greer received an anesthetic with chloroform, which had been introduced just the year before, and she died of a sudden cardiac arrest. Hannah's death sparked a public outcry and prompted a national inquiry into drug safety in anesthesia in the UK.[9] In subsequent years, awareness grew of the dangers of pharmaceuticals, and by 1881, a medi-cal book appeared devoted entirely to adverse drug effects, *The Untoward Effects of Drugs*, published first in Germany, then with a US edition only two years later.

Around the same time in the UK, physicians led by Ernest Hart, then editor of the *British Medical Journal*, were campaigning against the marketing of speciously patented medicines that contained poisonous or useless ingredients, but an early incarnation of the pharmaceutical industry defeated them: the Society of Chemists and Druggists opposed the physicians and lobbied effectively against an 1884 parliamentary bill that would have outlawed the sale of such concoctions, so that the bill did not pass. In the US not long after, Samuel Adams had more success, publishing a series of articles entitled 'The Great American Fraud' to lay bare the many false claims of patent medications and raise controversy over their unregulated dangers. His drawing attention to the issue led to the Pure Food and Drugs Act of 1906, which in turn led to the establishment of the

Food and Drug Administration (the FDA), which met with one of its early successes by exposing the useless concoction known as 'snake oil' as a fraud.[10]

Adverse reactions became more frequent and severe over the early years of the 20th century, as pharmaceutical companies introduced increas-ing numbers of potent drugs. Ill effects from the first antibiotics like sulphonamides and penicil-lin helped to raise awareness, but knowledge remained scattershot. Then a doctor's personal experience had an important result. Not long after World War Two, a Dutch physician became deaf, and the apparent cause was the antibiotics he received for tuberculosis, after the Nazi occupation of the Netherlands ended. Recognizing the need for a physician's reference source on the unintended effects of drugs and finding none, he began the research to create one himself. *Meyler's Side Effects of Drugs* was first published in English in 1952,[11] followed by multiple volumes every two to four years, each continuing to catalogue the unwanted effects of medications.

By 1973, after eight published editions of Meyler's book, the incidence of adverse events had grown so frequent that cataloguing began to require annual additions, each surveying the year's literature of pub-lished ill effects linked to pharmaceuticals. Today, encyclopedic editions are available; the 16th Edition of *Meyler's Side Effects of Drugs* expanded to seven volumes, and the total series now runs to thirty-three, plus an additional eight volumes dealing with specialty areas such as cardiology, cancer, and psy-chiatry.[12] These tomes, familiar to the 20th-century doctor, list types of medication and, for each, the pattern of their known adverse effects. Another resource structures the information differently: David Davis's *Textbook of Adverse Drug Reactions* is organized by types of adverse reactions and then identifies the drugs most commonly associated with each reaction.[13]

More recently, searchable databanks about possible adverse reactions to drugs have moved into electronic formats, but such resources are only as effective as the information entered into them, and without adequate tracking by pharmacies, doctors, and hospitals, the quality and currency of such databases is not yet keeping pace with the drugs in the marketplace. For the data to be useful, both in crises and in prescribing, the keepers of such data will also need to find effective means of tagging and organizing it – the groupings, patterns, and symmetries of and between drugs, across their effects, both the substances approved by the FDA and those which remain unregulated. Some of the progress in technology is in apps marketed directly to interested patients for monitoring their potential medications' side effects.[14]

Between 1952 and 1960, nearly 1,000 new medicinal products came to market, 118 of which were entirely new substances. Only after the thalidomide disaster of 1961 in Europe drew attention to the problem of inadequately tested drugs was serious attention given to adequate safety testing. To ensure the early detection of serious adverse effects, the FDA finally established that doctors and pharmacists need to be aware of known and potential issues, and that ongoing regulation of the pharmaceutical industry is necessary to keep doctors informed of adverse reactions.[15] Unfortunately, the current movement among lawmakers is in the opposite direction, towards less oversight and more leaving the process in the hands of those whose primary interests are profits rather than proven safety or effectiveness.[16,17]

Among adverse effects from prescription drugs, abuse is a particular tragedy, now an epidemic in the United States, in spite of legislative efforts to control them. Not a new problem, addiction has been a perennial challenge since the early days of opium trafficking, with federal legislation first introduced to make opium illegal in the early 20th century.[18] The Controlled Substances Act of 1970 was the most comprehensive federal law concerning drugs, creating a single system to regulate narcotics and psychotropic drugs[19] and establishing a legal framework that led to the 1973 creation of the Drug Enforcement Administration (DEA).[20] But abuse and addiction are rampant. The dispensing of prescription opioids increased 48% between 2000 and 2009,[21] and these drugs play a major role in the toll of unintentional deaths. Other than motor vehicle accidents, accidental poisoning with prescription drugs is the leading direct cause of unintentional death in the US, greater even than deaths from cocaine and heroin.[22]

That prescription drugs could cause so much suffering and death is a reflection of a failure to see the big picture about their production and distribution, when the DEA has the authority to regulate every step of the supply chain for these controlled substances, from the manufacturers to the distributors, prescribers and dispensers. A sizeable government agency, the DEA employs 5,000 special agents and has an annual budget of more than $2 billion. The agency is responsible for policing the dispensing of controlled substances, including those prescribed by doctors. The DEA must issue licenses to physicians who will prescribe narcotics, and the doctors must register with the DEA and pay a fee to prescribe these drugs.[23] Nevertheless, prescription opioids have become a major cause of death in the US and Europe.[24]

The existing systems for stemming the fatal tide of tragedy from opioids are proving inadequate, and clearly any solutions with a strong hope of success will take a multisystem perspective, looking upstream and downstream, considering precedent challenges of addiction, and involving multiple parties in the chain. From laboratories to the clinics, black markets to the street corners, insurance, regulation, enforcement and government, traditional and complementary medicine and education. And as the guardians of prescription drugs, physicians themselves can play a key role in limiting abuse or misdirected use of opioids, if they do not leave this problem of drug addiction to ostensible experts in other fields. They must also resist the unethical pressures of pharmaceutical companies to prescribe inappropriately, as demonstrated in current reports of opioid manufacturing executives pressuring and even paying doctors and nurses to prescribe a newest formulation off-label, for profit.[25,26]

We will always need opiates to relieve patients' pain, and there will continue to be multiple circumstances when their use is appropriate for relieving suffering. But indiscriminate and unmonitored use will always cause tragic problems, as we know from history, because there is no such thing as a non-addictive opiate, so they continue to wreak some of our most profound adverse effects.

Caring for safety or racing for profit?

If we apply the symmetry-seeking principles that prompt us to back up and consider the big picture, bigger than any one individual or disease, we unfortunately find a near-perfect mismatch – an asymmetry – in the vested interests of pharmaceutical companies and the desire for safe medicines, because their window for greatest profit is patients' window of greatest risk: the years immediately after a drug is first introduced. Even more unfortunately, those risks are not, in sum, counterbalanced with compensatory advantages.

New drugs, like other brand-new products, arrive on the scene with great hope, hype, and marketing fanfare, but in fact very few new drugs provide any significant advantage over the better-known medications. Experienced physicians prefer to use the better-known, tried-and-true therapies and suggest that most patients avoid trying the newest drugs for at least the first seven years of their general use, to allow time for the true profile of their benefit and harm to reveal itself. Independent studies report that in the last 30 years, fewer than one in every six new drugs offered significant advantage compared to the better-known alternatives. Among the approved and patented, most new drugs are more expensive than older drugs, but the data reveal what wellness professionals know from experience: that 85–90% turn out to offer little or no benefit over the better-known agents that experts prefer.

Nevertheless, these new, high-cost (because newly patented) pharmaceuticals account for an estimated $70 billion on the US drug bill, largely wasted on new minor-variant medications that offer trivial or no benefits and are not truly innovative and helpful.

With new medicines also comes the possibility – indeed, the likelihood – of new side effects, many of them both unexpected and adverse. As pharmaceutical companies in search of new patents tweak older medications, the new agents are unlikely to add benefit, so the balance of odds actually tips towards lesser benefit, as well as towards the prospect of greater harm. Anyone who takes the newest drugs is volunteering to be a guinea pig for the inevitably unfinished data collection on their side effects, even though the patient is apt to be an unwitting test subject, simply filling a prescription and following doctor's orders. In the meantime, pharmaceutical companies' financial interests are directly at odds with public health interests on this front, as their newly patented drugs are their most profitable – if they succeed in the market. Drug companies therefore advertise and promote their newest prescription drugs with glowing vignettes that fail to emphasize their true risk profile. By law, such adverts must include some specific warnings, but amidst the ultra-fast talk they don't mention the global reality that 1 in 5 of all new drugs will cause serious adverse reactions.

Over the past two decades, direct-to-consumer advertising of pharmaceuticals has aggravated the problem of drug overuse, misuse, and companion rise in adverse reactions. Such commercials were once strictly limited, but – under pressure from Big Pharma – the FDA eased its restrictions and opened the floodgates in 1997. Prime-time television is now torturous, with its endless repetition of the same pharmaceutical ads – which may describe bodily woes in intimate detail but offer virtually no useful information about drugs' actions. Still, the ads push for sales of ever-more expensive therapies and insist that you 'ask your doctor' about a company's newly patented product. Listen up for their closing babble of qualifiers, which include only a partial list of the known risks of adverse events. They do mention risk of death, almost always, but it's a warning guaranteed to be ignored or minimized when heard next to the promise of a wonder drug that consumers assume to be safe, if it's being sold in pharmacies and allowed in television ads. Last year, $5.2 billion was spent on direct-to-consumer drug advertising in the US – a practice that is not permitted in other countries, with only two exceptions (New Zealand and Brazil). The American Medical Association has lobbied for the abandonment of this direct-to-consumer advertising of pharmaceuticals, without success.

Meanwhile, audits of US hospital records reveal that 1.9 million hospital admissions are the result of adverse effects of prescription drugs, and this figure does not include admissions related to overdoses of prescription drugs, nor to inappropriate use of over-the-counter medications. And among patients already admitted to a hospital, a further 840,000 are estimated to suffer severe adverse reactions due to prescribed medication, making a total of over 2.74 million serious adverse drug reactions every year. This number includes only those that pass through our hospitals, not outside them, and only those events that are recognized as drug reactions. Of this number, 128,000 patients die as a result of these adverse prescription drug reactions, this death rate from medication ranking fourth (alongside stroke) as a leading cause of death in the US. Similar reports and audits by the European Commission reflect a similar experience, with an estimated 200,000 deaths in Europe, bringing the annual toll from severe adverse reactions to prescription medication to 328,000 deaths in the US and Europe together. [27-29]

At least 170 million Americans take prescription medications, and many derive significant benefits. Most drugs – about 80 percent – are generic, so they are well-established preparations with well-known patterns of risk and benefit. But, influenced by direct-to-patient advertising, millions of patients elect to take new, patented drugs that the

ads promote, even appearing in their doctor's office specifically to request them, as is the whole point of the advertising. Unaware that these new agents are the least well-known and the least well-tested chemicals for use in a diverse population, patients have seen only the marketed promise. In talking to patients, I compare taking a drug into your body to taking a lodger into your home; it is always much safer to choose someone you have known for a long while, rather than inviting in a complete stranger of unknown character and unpredictable behaviors.

If we compare hospital records to the financial balance sheets, we see that adverse drug effects have risen in tandem with industries' soaring profits. As aggressive pharmaceutical marketing practices have marked the last 30 years, and aggressive bacteria have raised their resistance to antibiotics, the general epidemic of adverse drug effects has mirrored the same pattern – soaring. And the epidemic is not limited only to hospitals: for every severe adverse event that prompts hospital admission, estimates suggest about thirty events of a less serious nature; this would represent about 81 million adverse reactions to prescription medications among the 170 million Americans under the doctors' care. And the greatest burden of both side effects and costs falls to those taking multiple prescriptions, including and especially the elderly.

The pharmaceutical industry has enjoyed ever-rising sales and soaring profits, despite a poor overall record of contributing to health or healing. Standout drugs that offer miracle cures are extraordinary but rare, and even some drugs touted for their early successes often prove useless or even dangerous after a while, though their exits are rarely as loud as their entrances were – there are no ads during the Super Bowl about drugs that were quietly withdrawn from the market after they were proven to cause stroke or bleeding or heart attacks.

A recent published review from an expert panel in France reported that only 15 of 994 new drug products introduced between 2002 and 2011 offered true therapeutic advantage, and a further 61 of 994 were considered to offer some advantage,[30] but according to Marc-André Gagnon's research, the pharmaceutical industry has enjoyed a nearly 200 percent gain in net returns on their revenues over recent decades: profits rose from a healthy 10% in the 1970s to 16% by 2000, and to 19% by 2010. Compared to other major Fortune 500 companies, which average a net return of 5-6%, the return on revenue for pharmaceutical companies runs from double to more than triple the usual return of other corporations.[31] And disproportionate amounts of their operating budgets go to marketing, compared to what they spend on actually developing and manufacturing their drugs.[32]

Gagnon identified two factors that have driven drug companies' extraordinary profits: ever-rising prices and increased drug prescribing. The two are linked, because the necessity of prescriptions is one of several factors that skews prices in the drug market. Symmetry-seeking reveals the mismatch of medical commerce, where the doctor places your order, but does not have to pay. Like overpriced textbooks that students would not have chosen to buy but must because their teacher assigned them, the only prescription drug a patient can take is the one a doctor endorses. Much of the runaway cost of 'health care' stems from this kind of disconnect, in contrast to nature's biotensegrity architecture where everything is connected, always linked together as balance and counterbalance.

Patients pay dearly in a free market

There is no easy, one-industry fix for the imbalances in our interlocking system of drug production, distribution, and consumption that will ensure broad improvement in safety and effectiveness, so solutions require an eye for inter-systems symmetry – looking for imbalance and counterbalances. Certainly, health care professionals with the power to prescribe and dispense medications will continue to play a critical role in addressing the problems that stem from misuse and overuse of pharmaceuticals. We continue to need their expertise, based on years of medical training, to understand what drugs are effective and which are safe, in what circumstances. But doctors and nurses themselves clearly need more up-to-date training, based on contemporary data, and we all need much better tools and accurate, open-access resources for prescribers, pharmacists, and patients alike. And those better data will rely on a complex of collection and analysis from multiple contributors, including public health organizations, pharmacies and pharmacists, hospitals, drug companies, farmers and industrial agribusinesses, and regulatory agencies – as well as others we may have overlooked.

An effective FDA is a necessary linchpin in any effort to face problems with pharmaceuticals in the US. It was built on laudable principles, to face precisely the sorts of difficulties now before us – publicly funded, requiring independent reviewers and regulators – but in recent years Congress has relinquished responsibility for funding drug approval to the pharmaceutical industry. Industry-funded studies provide less-than-impartial data and focus more on streamlining the process to bring new drugs to market quickly rather than on making sure drugs are both effective and safe. Research on the contemporary FDA now describes it as an extension of pharmaceutical industry, playing a significant role in expanding markets so that ever-more people will take ever-more drugs[33] – not protecting American lives and health. Strengthening their ability to act independently and in the public interest will be crucial to success at reining in abuses.

When tackling these dire challenges with hormones, antibiotics, opioids, and other drugs that have wide negative impacts on our health, we can also draw our axes of symmetry between problems that have shown similar patterns of success and failure, such as cigarette smoking, history's opium-taking epidemics, or measures that have and have not been effective in other countries on the same or similar problems, or in smaller communities. What will scale to meet our broader epidemic?

Ultimately, the most important contributor to a better use of pharmaceuticals is you – both as a consumer and as a citizen. Continuing what you have learned here and elsewhere, you can learn to ask ever-better questions of your doctor, your pharmacist, and all your health advisors about their pharmaceutical prescriptions and recommendations – about safety, effectiveness, cost, sources, and side effects. You can approach new-to-market drugs with the appropriate care and skepticism. You can ask for (and heed) guidance for non-pharmaceutical self-interventions – the lifestyle changes that are most likely to contribute to your long-term health and wellness. And, remembering that you are the expert on your own body, you can pay attention to your own reactions to drugs so that you can be a full partner in decisions about what works, because each of us individually will enjoy the benefits or suffer the dangers of whatever drug interactions and adverse reactions pharmaceuticals bring.

We can also seek out foods without hormones, antibiotics, and other chemical preparations and problematic feeding practices in their growing or processing, while pressing for best practices and for transparency in farming practices, which are the topics of the next two chapters.

References

1. Penicillin finder assays its future. *New York Times*, 26 June 1945: 21.
2. Centers for Disease Control and Prevention (2013). Antibiotic/Antimicrobial Resistance, Biggest Threats. Online: https://www.cdc.gov/drugresistance/biggest_threats.html.
3. Kar A (2015). Antibiotic use in livestock going up, up, up. Online: https://www.nrdc.org/experts/avinash-kar/antibiotic-use-livestock-going.
4. Kesselheim A, Avorn J, Sarpatwari A (2016). The High Cost of Prescription Drugs in US: Origins and Prospects for Reform. *Journal of the American Medical Association*, 316(8): 858–871.
5. IQVIA Institute for Human Data Science (2017). Medicines Use and Spending in the U.S.: A Review of 2016 and Outlook to 2021. Online: https://www.iqvia.com/institute/reports/medicines-use-and-spending-in-the-us-a-review-of-2016.
6. The Pew Charitable Trusts (2018). A Look at Drug Spending in the U.S. Estimates and projections from various stakeholders. Online: http://www.pewtrusts.org/en/research-and-analysis/fact-sheets/2018/02/a-look-at-drug-spending-in-the-us.
7. AMA (2015). AMA Calls for Ban on DTC Ads of Prescription Drugs and Medical Devices. Online: https://www.ama-assn.org/content/ama-calls-ban-direct-consumer-advertising-prescription-drugs-and-medical-devices.
8. Blake V (2013). Fighting Prescription Drug Abuse with Federal and State Law. Online: http://journalofethics.ama-assn.org/2013/05/hlaw1-1305.html.
9. Haggard WD (1908). Chloroform anesthesia. Online: https://jamanetwork.com/journals/jama/article-abstract/428415?redirect=true.
10. Gandhi L (2013). A History Of 'Snake Oil Salesmen'. Online: http://www.npr.org/sections/codeswitch/2013/08/26/215761377/a-history-of-snake-oil-salesmen.
11. Meyler L, Herxheimer A (1952). Side Effects of Drugs. Amsterdam: Excerpta Medica.
12. Aronson JK (Editor) (2016). Meyler's Side Effects of Drugs: The International Encyclopedia of Adverse Drug Reactions and Interactions (16th edition). Amsterdam: Elsevier.
13. Davies DM, Ferner RE, de Glanville H (1998). Davies Textbook of Adverse Drug Reactions (5th edition). Chapman and Hall Medical.
14. Tinari G (2016). 3 iPhone Apps to Make Life Easier Tracking Your Medications. Online: https://www.guidingtech.com/58980/medication-tracking-ios-apps/.
15. FDA. Development & Approval Process (Drugs). Online: https://www.fda.gov/drugs/developmentapprovalprocess/.
16. Lilleston TR (2017). Drug Approval Process Far From Foolproof. Online: https://www.aarp.org/health/drugs-supplements/info-2017/fda-drug-approval-safety-risks-fd.html.
17. Kaplan S (2017). Trump derides 'slow and burdensome' approval process at FDA. Online: https://www.statnews.com/2017/02/28/trump-address-rare-disease-drugs/.
18. Hohenstein K (2001). Just what the doctor ordered: the Harrison Anti-Narcotic Act, the Supreme Court, and the federal regulation of medical practice, 1915-1919. *Journal of Supreme Court History*, 26(3): 231.
19. US Act of Congress (1970). Controlled Substances Act, US Code 21, Chapter 13, Section 801 et seq.
20. Drug Enforcement Administration. The DEA Years,1970-1975. Online: https://www.dea.gov/about/history/1970-1975%20p%2030-39.pdf.
21. Office of National Drug Control Policy, US Executive Office of the President of the United States of America (2011). Epidemic: Responding to America's Prescription Drug Abuse Crisis. Online: https://www.ncjrs.gov/App/Publications/abstract.aspx?ID=256103.
22. Centers for Disease Control and Prevention. Opioid overdose. Online: https://www.cdc.gov/drugoverdose/index.html.
23. Drug Enforcement Administration. DEA history. Online: https://www.dea.gov/about/history.shtml.
24. Bugge A (2017). Drug deaths on the rise in Europe for third year: report. Online: https://www.reuters.com/article/us-europe-drugs/drug-deaths-on-the-rise-in-europe-for-third-year-report-idUSKBN18X1Y4.
25. Dwyer C (2017). Pharmaceutical Founder Arrested in Alleged Nationwide Opioid Scheme. Online: http://www.npr.org/sections/thetwo-way/2017/10/26/560263997/pharmaceutical-founder-arrested-in-alleged-nationwide-opioid-scheme.

26. Snow A (2017). Billionaire drug company founder charged with bribing doctors to prescribe opioids. Online: http://www.chicagotribune.com/business/ct-biz-insys-john-kapoor-pharma-drug-bribery-20171027-story.html.

27. Bouvy JC, De Bruin ML, Koopmanschap MA (2015). Epidemiology of Adverse Drug Reactions in Europe: A Review of Recent Observational Studies. *Drug Safety*, 38(5): 437–453.

28. Stausberg J (2014). International prevalence of adverse drug events in hospitals: an analysis of routine data from England, Germany, and the USA. Online: bmchealthservres.biomedcentral.com/articles/10.1186/1472-6963-14-125.

29. Angiji A (undated). Adverse Drug Reactions related to mortality and morbidity: Drug-drug interactions and overdoses. Online: www.xendo.com/images/pdf/ADRs-by-Major-systems-II-Final-Report.pdf.

30. Prescrire International (2011). New drugs and indications in 2011. France is better focused on patients' interests after the Mediator scandal, but stagnation elsewhere. Online: http://english.prescrire.org/en/115/445/48384/2384/2383/SubReportDetails.aspx.

31. Gagnon MA (2013). Corruption of pharmaceutical markets: addressing the misalignment of financial incentives and public health. *The Journal of Law, Medicine & Ethics*, 41(3): 571–580.

32. Lauzon L-P, Hasbani M (2006). Analyse socio-économique: industrie pharmaceutique mondiale pour la période de dix ans 1996–2005. Montreal: Chaire d'études socio-économiques de l'UQAM. Online: https://unites.uqam.ca/cese/pdf/rec_06_industrie_pharma.pdf.

33. Light DW (2013). Risky Drugs: Why The FDA Cannot Be Trusted. Online: https://ethics.harvard.edu/blog/risky-drugs-why-fda-cannot-be-trusted.

Fueling the fire: downstream drift

On April 19, 1995, a young army veteran named Timothy McVeigh drove a rental truck filled with fertilizer into Oklahoma City. When he detonated this load of ammonium nitrate outside the Alfred P. Murrah Federal building, the blast was – and remains – the most destructive act of domestic terrorism in US history, claiming 168 lives, injuring more than 680 other victims, and destroying or damaging 324 buildings.

The fertilizer he used in the attack has a long history of killing, including America's most deadly industrial disaster. In 1947, a ship exploded in Texas City, Texas, destroying the port where it was docked, knocking two planes out of the sky, and killing almost 600 people.[1] We have yet to learn how to manage this fertilizer's volatility, so the danger continues to the present day. When a fertilizer storage facility in the Texas town of West caught fire on April 17, 2013, nearly two decades after the Oklahoma bombing, the exploding ammonium nitrate destroyed 80 homes and a school, killing fifteen and injuring another 200. Because of the explosive character of ammonium nitrate fertilizer, known to be lethal, the warehouse had presented a catastrophe waiting to happen – this time, caused by arson.[2]

These lethal explosions continue to occur, and not only in the US. In China, where fireworks were first invented more than two thousand years ago, a warehouse fire in the seaport of Tianjin detonated a load of ammonium nitrate fertilizer in August 2015 that killed 165 and injured 800 more, destroying over 300 buildings and a railway station in the blast and causing more than $1 billion in damage.

Ammonium nitrate is powerful stuff.

Fireworks and fertilizer

It is hard to believe a familiar compound that farmers use to grow plants could have anything in common with lethal explosives, but it is no coincidence that fertilizer is a terrorist's explosive of choice. The answer lies in chemistry, with the element nitrogen. For farming, fertilizer is spread wide and soaked with water, to release nitrogen slowly and feed the plants, but when concentrated and lit with fire, the result is a chemical reaction in which gases expand and temperatures soar instantaneously, releasing intense heat and pressure – in other words, an explosion.

The ancient Chinese created the first fireworks using a naturally occurring white salt that captured the energy of nitrogen as its explosive charge: sodium nitrate, called 'China snow'. Centuries later, in 1846, Italian chemist Ascanio Sobrero invented the first synthetic nitrogen-based explosive, nitroglycerin, which proved too dangerous and difficult to manage, being so highly and spontaneously volatile. But Swedish chemist Alfred Nobel soon discovered that mixing sand with the nitroglycerin rendered it more stable. Choosing the name 'dynamite', he patented the compound, along with its detonators and other explosive paraphernalia. Then he built the first factories to manufacture nitroglycerin and dynamite, giving a great gift to civil engineering and to the mining industry but also creating a dreadful scourge of killing power, one that would fuel military conflicts and warmongering for decades to come. Nobel's inventions brought him fantastic personal wealth, but he was also aware that he had let a destructive genie out of the bottle, and before his death in 1896, he created the Nobel endowment to honor achievements in the services of peace and humanity, literature, and science.

German chemist Fritz Haber won Nobel's prize in 1918, as the first to master an effective method of fixing nitrogen. As told in the extraordinary book *The Alchemy of Air: A Jewish Genius, A Doomed Tycoon, and the Scientific Discovery that Fed the World, but Fueled the Rise of Hitler*, Haber found a simple way to combine the nitrogen freely available in the air with hydrogen, producing ammonia, one of the essential ingredients in ammonium nitrate fertilizer.[3] His discovery would eventually feed an exponential increase in crop yield around the globe, but, like Nobel's dynamite, it also fueled massive destruction in the ongoing wars, which enlisted the highly explosive ammonium nitrate in their bombs. Its first major commercial accident was a deadly factory explosion in Oppau, Germany, in 1921, where a blast at an ammonium nitrate plant leveled the town and killed almost 600 people, injuring thousands more.

Over the 20th century, the primary use of ammonium nitrate alternated between farm fertilizer and war explosives. On entering World War One in 1917, the US needed munitions immediately, so the government commissioned two giant industrial complexes to be built near the town of Muscle Shoals, Alabama, chosen for its river's promise of hydroelectric power. These munitions plants were to be the largest in the world, and in 1918 construction began on a dam spanning the Tennessee River. The war ended almost as soon as construction began, but the dam building continued until it was finished, in 1924, long after the plants could be used to produce explosives for the war. Henry Ford offered to buy the site, but after years of debate over what to do with the $130 million complex, Congress finally decided

in 1933 to create the Tennessee Valley Authority, an entity consolidating multiple resources towards multiple purposes as part of its mandate: building dams and lock gates to control flooding in the lower Mississippi valley and creating deep-water navigation, as well as managing experimental chemical plants. When World War Two came soon afterwards, it brought further expansion and investment in the plants, including increased production of the synthetic ammonium nitrate now needed again for the war effort.

After a second war, the plant's production capacity was well-established and in search of a new market, of peace-time consumers. To continue to manufacture in these same commercial facilities in Alabama and elsewhere, the military–industrial complex therefore redirected its focus, repackaging ammonium nitrate to market it for civilian use as synthetic fertilizer, shifting the product line from wartime armaments to domestic agriculture. By the end of the war in 1945, there were 10 synthetic nitrate factories in the US. In Germany, Europe, and Japan, factories had been destroyed, so the US had the dominant position to promote and distribute a new product. Renamed the National Fertilizer Development Center, the Alabama munitions factories thus found new life, in an apparent win-win-win decision for the US government, industry, and agriculture. Fertilizer created a new market for utilizing the postwar stockpiles of synthetic ammonium nitrate – a global market, for a new product that promised improved agricultural production not only for the nation but for the planet, in a world hungry for growth after years of war and deprivation. Only now, what would blow up was not only dangerous stockpiles of fertilizer, whether in Texas or Oklahoma or China, but our waistlines, across America and, eventually, around the world.

Driven by heavy-handed industrial interests and obvious short-term gains, this chemistry experiment with ammonium nitrate played out on a massive scale. With limited understanding or regard for possible long-term consequences, decision-makers promoting this synthetic fertilizer set a national and worldwide course that would have a profound impact on land use, farming practices, ecosystems, and our shared human fate.

Compost or chemicals?

The symmetry-seeking principles shift our perspective to a wider and longer view of the ammonium nitrate revolution in farming, revealing what those decision-makers failed to consider – and still resist facing today. Beginning at the most familiar human scale, we find that even the home gardener has absorbed the fundamental chemistry lessons that undergird best practices in farming, since gardening is essentially small-scale farming. In both, traditional principles of working the land have evolved over tens of thousands of years, and the first fundamental tenet of good gardening is to look after the dirt. An experienced gardener feeds and nurtures the soil and then can leave it to the soil to feed the plants. Composting recycles organic waste from the kitchen and from the garden itself, making all its nutrients and micronutrients available to plants through bacterial decomposition. The same heat and moisture that drive microbial growth also spur plants to grow, providing nutrients as they're needed, in step with plant growth. Compost adds organic bulk to the soil, which helps to improve the size of soil particles and thus water retention and porosity, thereby improving drainage. Compost also helps to balance soil pH away from too much acidity and renders the nutrients already in the soil more readily accessible to the roots of growing plants.

Fall is the season when the experienced gardener will swathe the beds with compost, giving the fibrous materials time to decompose, hold moisture, and build soil structure while the nutrients, minerals, and micronutrients prepare to supply all the nourishment the plants will need for the coming year. Cultivating the soil is the first order of business for the gardener, just as nurturing the whole patient with a view to next year and the next, preparing for the years ahead, is the charge of the skilled physician.

Specific crops vary in their demands, drawing particular nutrients and mineral resources from the land, and crop rotation is the deliberate practice of growing different plants in the same soil bed in different years and seasons, to help balance the soil and avoid depletions. A plant needs a whole host of building bricks and bits and bobs if it is to build and grow to its full potential, a complete, healthy, fully formed specimen bearing a full seedhead. The finest yields come from fresh land not previously farmed, in virgin soil not depleted of any nutrients. And the opposite of crop rotation is monoculture, growing the same crop year after year in the same fields, which will deplete the soil and eventually deprive the plants of essential nutrients, micronutrients, and minerals, with diminished crop yields, more vulnerability to disease, and increased risk of partial or complete crop failure.

Fertilizers attempt to replenish the soil to replicate the richness of virgin soil, and experienced gardeners favor compost over chemical products because compost feeds all of the soil's complex needs, whereas 'plant food' does not. A familiar commercial product such as Scott's Miracle-Gro is a '15-30-15' fertilizer – a simple chemical concentrate of the three main nutrients needed by plants: 15% nitrate, 30% phosphate, and 15% potassium. With these three ingredients, Miracle-Gro does often produce miraculous results, and it is easy to use – just dissolve the neon blue powder in your watering can, and you are off. Many other fertilizers are equally effective and convenient, but the Miracle-Gro brand has become synonymous with fertilizer, because its true miracle is effective product naming, advertising, marketing, labeling, and display – always on sale, of course, placed prominently in the garden store.

But a diet too rich in some nutrients and starved in others is just as harmful to plants as to humans, it turns out, so the current industrial-scale applications of synthetic fertilizer products are problematic.

Like composting, which returns unused plant parts back to the soil where they were grown, the traditional practice of tilling after a harvest turns the remaining vegetable matter lying on and in the ground back into the land, recycling all the nitrogenous leftovers from last year's growth. Some vegetables such as legumes, peas, beans, and alfalfa also have nitrogen-fixing bacteria in their roots, so when planted in a crop rotation, they actively replenish soil that has been depleted of nitrogen. This mutually beneficial arrangement between the root bacteria and the plants illustrates one of many counterbalancing 'better together' partnerships in nature – symbiosis. In exchange for providing nitrates to the plants, the bacteria receive safe sanctuary, fluids, sugars, and carbohydrates in return.

Fixed nitrogen available in the soil is the nutrient that almost always sets the limits on a plant's rate of growth, because few plants can take any from the tightly bound paired nitrogen atoms in the atmosphere. Despite nitrogen's abundance in the air we breathe, plants must depend on the limited supply of water-soluble nitrate salts available in the soil. The central problem with synthetic ammonium nitrate is its high-energy double-dose 'nitrate-nitrate' structure. Its components each contain one nitrogen atom, with ammonia made of nitrogen and hydrogen (NH_3) and nitrate made of nitrogen and oxygen (NO_3). But together in ammonium nitrate, they bring a double dose of nitrogen, making for an unstable mix of this volatile element: NH_4NO_3. By comparison, the naturally occurring saltpeter (sodium nitrate) such as found in the first fireworks, holds only one nitrogen atom, therefore lacking the hotheaded nature of ammonium nitrate.

Farms too big to fail

In traditional agriculture, the successful centuries-old method of cultivation is 'mixed land use', which depends not only on crop rotation but on growing crops and raising animals together on the same land. Animals get what they need for growth by feeding on plants, living or dead, and the animals perpetuate a cycle of balance by returning their nitrogenous waste in the form of urine and manure to the pasture. Dust to dust and soil to soil. These centuries-tested methods of composting, tilling, crop rotation, and mixed land use sustain an optimum balance of nitrogen, among many other nutrients, over years and years of cultivation.

During the same period that the chemicals we discuss in this chapter first began to appear in farming – but before vast industrial agribusiness fully took hold – other forces were laying the roots of today's problematic practices, in understandable efforts to increase crop yields. Mechanical tools such as tractors transformed a farmer's ability to till larger swaths of land, and efficiencies of scale lent themselves to wider planting of single crops, row upon row, acre upon acre. The threat from this wide monoculture then made itself felt early on, far beyond losses of crop size or crop quality.

Collective memory is short, but America has lived through a well-known disaster of failed land policies already, in which monoculture led to erosion on a scale that provoked famine and national economic disaster, displacing huge populations. After World War One, rising wheat prices and increased demand from Europe encouraged farmers to plow up millions of acres of native grassland to plant wheat. Favorable conditions in the Great Plains and good yields expanded ever-greater confidence, but then, in 1931, extreme weather conditions and drought set in. Crops began to fail, exposing the bare farmland, with no live roots or ground cover to hold it in place. The soil began to blow away, leading to massive dust storms across the US and deepening the Great Depression of the 1930s.[4]

As land conditions continued to deteriorate, the worst day of weather arrived on April 14, 1935, 'Black Sunday', when a wall of blowing sand and dust started to move east from the Oklahoma Panhandle. An estimated three million tons of topsoil was blown off the Great Plains in a single day. Altogether, roughly 2.5 million people fled the 'Dust Bowl' states of Oklahoma, Nebraska, Kansas, Texas, and New Mexico in the 1930s – the largest migration in US history. Oklahoma was worst hit, losing 440,000 people to migration. Many were poverty-stricken refugees who moved to California seeking work and living in shantytowns. John Steinbeck memorialized the plight of these 'Okies' in his 1939 novel *The Grapes of Wrath*. Today's extreme weather conditions of superstorms and flooding, extended drought, and wildfire – all amidst climate change – highlight the dangers of relying on monoculture across wide swaths of land, from year to year to year in the same fields.

Amidst the shifting mores of farming over the 20th century, nitrate fertilizers had an impressively positive impact on harvest size and a crop's immediate productivity, and they continue to fuel that growth today. By reducing the need to take time off from the main crop before a next planting, whether by crop rotations or by leaving fields fallow, they increase yield per acre. This productivity has even made it possible to reduce the area of land devoted to farming, in some places, allowing for the expansion of forests, grasslands, and wetlands, while simultaneously feeding a growing global population. And for soil that is already out of balance from intensive cultivation and monoculture, nitrate fertilizer has helped to prevent further depletion of nitrogen stores and made it possible to plant more cash crops and staple foods, instead of forage legumes such as alfalfa and clovers, which farmers have historically needed to plant in crop rotations, to maintain soil fertility and fix nitrogen in the soil. But monoculture generally and synthetic nitrate fertilizer particularly throw multiple systems out of balance, to the detriment of human health and sustainable land use.

Earth chemistry 101

Naturally occurring saltpeter, the China snow of the ancients, and seabird guano from Peru are each high in nitrogen and useful for fertilizer – as was the nitrate mined from Chile, highly prized until Europeans exhausted its supplies in the late 19th century. All these organic sources of nitrate come from the earth, unlike the nitrogen that fuels our bombs and synthetic fertilizer, once Haber learned to capture it from the air.

Our world consists of three great realms – land, sea, and air – and all the elements of the periodic table are represented and balanced within and between each of them. As if they were neighboring countries, they exchange solid, liquid, and gas at their shared borders, but each realm is distinct, because each holds contrasting proportions of the primary elements. The sea, which covers most of the earth's surface, is mainly water, of course – H_2O. Even though seawater also contains a host of salts and minerals in dilute proportions, the bulk of salt water is still simply hydrogen and oxygen. In contrast, the air is mainly nitrogen. We humans may think of oxygen as the critical element in air, but only about 20% of the air we breathe in is oxygen, while almost 80% of our inhaled and exhaled breath is nitrogen. And on dry land, the most plentiful element by far is carbon: the soil and all the trees and plants, animals, and other creatures growing on and from the land are rich in carbon and hydrogen.

As carbon is abundant on land and rare in the air, so the nitrogen abundant in air is relatively scarce in the soil. Despite its key role in the structure of all proteins, amino acids, and DNA in all organic matter – and although plentiful in the natural world as gas in the air we breathe – nitrogen and nitrogenous compounds are surprisingly sparse in soils and the oceans. On land, most nitrogen is contained in plants and animals. Living animals excrete some in urine and feces, and bacterial digestion and decomposition recycle dead plants and dead bodies to release nitrates and ammonium salts. But in healthy fertile soils, only small amounts of naturally occurring nitrate are present and available to be taken up through the roots to feed plant growth.

Hydrogen is the most common element in the universe. It is also the smallest, with only one proton and one electron (atomic number 1 on the periodic table of elements). Because it is light and nimble, hydrogen attaches readily to other molecules and to itself, and two of these identical atoms together (H_2) make

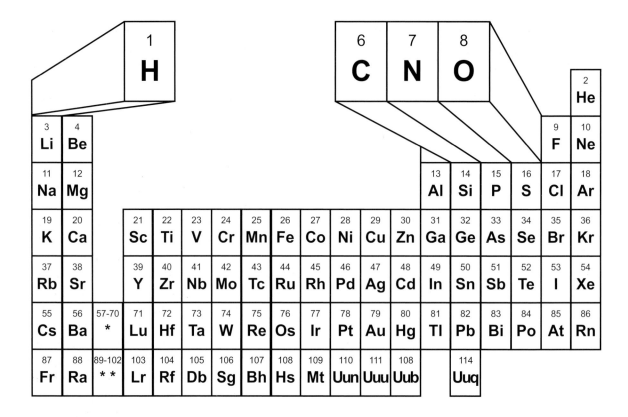

Figure 6.1

The periodic table of elements. Note how tightly clustered are carbon, nitrogen and oxygen – three immediate neighbors in the wide expanse of possible places in the table, as if the result of spinning in a great centrifuge. The like-weighted elements together with hydrogen make up the bulk of our living world of land, air and sea.

hydrogen gas. Carbon, nitrogen, and oxygen are immediate neighbors in the periodic table, with their atomic numbers 6, 7, and 8, respectively. Like hydrogen, nitrogen has an odd number of protons and electrons and exists predominantly as two atoms bound very strongly together, in the form of nitrogen gas (N_2), and like hydrogen, nitrogen's chemistry conveys its explosive value in bomb making.

Among all those different elements in the periodic table, with their distinct qualities and variation, it is staggering to realize that nature uses only a tiny cluster of three elements over and over again: carbon, nitrogen, and oxygen. Together with hydrogen, these three construct our natural realms of land, air, and sea. Here nature displays the same relentless economy as demonstrated in our DNA; the genes offer almost endless diversity and variation, but only the tiny cluster of Hox genes appears again and again in constructing life's living organisms, including our own.

At the boundary zones between these domains of land, air, and sea, constant interaction and agitation prevail. Think of waves crashing on a beach, pushing and pulling at the sand, broken shells washed this way and that while successive waves pound the ground, giving up energy from the water to the shoreline. At this meeting edge, the water's restless push-pull beats against the shore, disturbing and churning the sands. Solids wash forward and up the beach with the advancing waves, then fall back down in retreat. Not only here at the shoreline but also across the ocean floor, land and sea touch and interact. And throughout this vast contact interface, land and sea shape and reshape one another, attached, interdependent, and in constant interplay – as do sea and air, through precipitation, evaporation, and condensation.

Beyond these ribbon boundary zones, the terrestrial, marine, and atmospheric compartments remain distinct. When healthy, their environments rest in a dynamic balance and allow only limited exchange across the boundaries. On land, natural clouds of smoke and ash billow into the air from active volcanoes, similar to forest fires and industrial plumes that both send megatons of carbon and other elements up into the sky, together with particulate emissions to mix with the atmosphere. Smoke rising from fires is easy to see – even from small ones, such as chimneys and flues – but exhaust fumes from cars and trucks tend to be invisible, unless their smoke mixes with the morning fog and hangs over a cityscape as clouds of harmful smog, dense and persistent. In the same way, it is also impossible to discern the great clouds of methane gas expelled from the gut and fecal lagoons of the commercial hogs and cattle held in confined animal feedlots, or the fumes rising from city waste and garbage dumps. Notwithstanding the reality of these emissions, their invisibility makes them easy to ignore, even as we experience their impacts.

Much of the mixing between compartments is within the range of historical norms, and much of it is beyond our control, but human activity increases agitation and mixing across the boundaries. Inevitably, as our living consumes and creates, we shape the environment, as we are in turn shaped by its constraints. Absolute balance and harmony are romantic ideals, but humans – by our number and nature, and with our tools and activities – can push disproportions so far as to wreak havoc on this

world, causing our environment to deteriorate much more rapidly than it would without us, thereby limiting the sustainability of our own healthy human life on earth.

In the delicate balance of the primary elements between the air, sea, and land, hydrocarbons (made of hydrogen and carbon) are stored safely in the earth, deep under the land. Formed from ancient vegetation and then subjected to intense heat and pressure under the earth's surface over millennia, hydrocarbons are now coal, oil, and natural gas – the high-energy compounds that we use to power modern machines and call 'fossil fuels'. Concerns about human influence on global warming have centered on carbon emissions because burning too much fossil fuel risks transferring too much carbon out of the land compartment into the air, disturbing the natural balance and promoting climate change.

Unfortunately, the production process of the synthetic chemicals for farming deepens the problematic impact of current agricultural methods. Nitrogen available for fixing may float freely in the air, but manufacturing ammonium nitrate is expensive, demanding high temperatures and pressures to combine the ammonia and nitric acid, all of which requires a great deal of electrical energy, whether from hydroelectric power, natural gas, or petroleum. As Michael Pollan puts it in *The Omnivore's Dilemma*, 'Fixing nitrogen allowed the food chain to turn from the logic of biology and embrace the logic of industry. Instead of eating exclusively from the sun, humanity now began to sip petroleum'.[5] An estimated 50 gallons of oil are needed to grow an acre of corn. So carbon emissions are another dimension of the nitrate problem.

Downstream from corn

Adding carbon into the atmosphere is only one of the ways contemporary farming upsets the balance between earth's elemental boundaries, between land and air. Another is throwing the nitrogen balance dangerously askew between land and water.

Some plants and grasses have voracious appetites for nitrates and react to fertilizer with extraordinary growth. No plant is more responsive than Zea mays, the proper name for maize, also known as corn – a plant native to North America. The more nitrogen you feed to corn, the more magically it grows, and the cost of the fixed nitrogen that can be captured from the air and poured onto the soil is negligible compared to the profits from the prolific fertilized crops. With a little cross-breeding and genetic engineering, agribusiness has brought the fairy tale of 'Jack and the Beanstalk' to life. Records of farm production from the 1880s reflect no significant increase in corn yield until the introduction of synthetic nitrate fertilizers in the 1940s. But after producing about 39 bushels per acre on less than 10 million acres in 1935, corn yields rose to 62 bushels an acre in 1960. By 2000, Iowa averaged a massive 145 bushels per acre, across 12 million acres of corn.[6]

Popular products such as Miracle-Gro rely on nitrogen and phosphorous, plus added minerals like potassium, calcium, iron, and magnesium, to expand crop size and yields beyond their former limits – another triumph of commercial success, with human ingenuity expanding the harvest. What could be wrong with that? But, unlike traditional farming methods that feed the soil first and the plants second, synthetic chemical fertilizers take the shortcut of feeding today's yield at the expense of tomorrow's crop.

Nitrates are highly soluble and thus readily washed from the soil into creeks and rivers. Industrial agribusiness, or 'Big Ag', has therefore severely altered the balance between the natural compartments of land, sea, and air, especially through heavy use of nitrates in its industrial corn production. The synthetic nitrogen deliberately ladled onto cornfields has nowhere to go but down into rivers, lakes, and estuaries, then on to the coastal areas where runoff reaches the sea. Because all the major rivers of Iowa are part of the Mississippi River's watershed, the great Iowa cornfields in the upper Midwest drain to the south, where the river meets the ocean at the Gulf of Mexico. And as it is

for corn – more nitrate, more growth – so it is for other vegetation; too much nitrogen and nitrate runoff can cause nutrient loading in surface and marine waters, known as eutrophication. Simple plants that live in the river's fresh water or in the sea are stimulated to grow out of control. Colonies of algae then spread like oil spills on the surface of the water, reaching for hundreds of miles, their harmful blooms forming dense carpets that block the light and consume all available oxygen. Like bacterial overgrowth, which often produces toxins with harmful effects, the algae takes its toll on fish, shellfish, other marine life, and birds. Thus, through the magic of synthetic fertilizers, your sweetcorn can kill your crab cake.

Since the 1960s, when too much algae became a regular feature of US coastal water, marine biologists have come to refer to these areas as 'red tides' or 'dead zones', after the preponderance of fish and crab deaths there. But they are also called coastal 'blooms', and aptly so, after the thriving algae that flourishes there. Feeding is essential for any living thing, but too much is as dangerous as too little. A healthy habitat, like a healthy body, cycles in a balanced middle range. And unhealthy habitats are intrinsically related to unhealthy bodies: ammonium nitrate fertilizer feeds the corn that drives the yields that convert to syrup and fatten children, overwhelming the pancreas and provoking the complications of diabetes, heightening risks of early death. Meanwhile, in the habitat, excess fertilizer also spills over beyond the plants, escaping from the fields to the streams, rivers, and coastal waters, finally flooding the algae with unwanted feeding, all exploding a deadly glut of ugly overgrowth and unwelcome poisoning by toxic wastes.

In recent years, such blooms have been reported in almost every US coastal state and are becoming a national concern, due to their adverse impact not only on human health and marine ecosystems but also on the economic health of regional fisheries and tourist economies. A multidisciplinary panel of scientists has published a consensus report listing the major causes of harmful algal blooms and proposed strategies for trying to manage this growing threat.[7] They report that nitrogen, phosphate, potassium, and other fertilizer products used in agriculture are significant factors that contribute to the nutrient-rich conditions of fresh water runoff that serve to fuel downstream harmful algal blooms.

Other regions of the US and other countries have followed Iowa's lead and applied the same flawed industrial-style methods to boost corn yield per acre, creating the same patterns of downstream nitrate pollution and coastal blooms in Europe and Asia. The irresistible prospect of higher crop yields has not only driven copious overuse of synthetic nitrate fertilizers, but also propelled agribusiness towards monoculture, even though planting the same corn crop year after year is a profoundly flawed strategy that threatens soil quality, pushing it towards greater acidity and, finally, exhaustion. Still, worldwide, farmers currently apply more than a hundred million tons of synthetic nitrogen fertilizer every year.[8]

Drinking water woes

Overuse of nitrogen captured from the air, poured onto our plants, has even more direct effects on human health than through its persistent long-range damage to our oceans and all that lives there, and one effect is on our water supply. In a modern civilization with a well-established infrastructure, we should be able to rely on excellent drinking water, but in some communities, ranging from California to the New York islands, tap water is tainted with heavy nitrate pollution.[9] Though odorless and tasteless, nitrate is harmful to human health, implicated in respiratory and reproductive problems, thyroid disease, some cancers, and blue-baby syndrome, and scientists now report unacceptably high levels of nitrate contamination in much of our drinking water.

In one of the most intensively farmed areas in the world, the Central Valley of California, production methods rely heavily on irrigation and on synthetic fertilizer. Studies reveal increased nitrate pollution in the Salinas Valley and Tulare Lake Basin regions, and agriculture is thought to be responsible for 96% of the contamination, 54% of it from synthetic nitrate fertilizer runoff and 33% from animal manure. And on the east coast, suburban Suffolk County in Long Island, New York, suffers a similar difficulty, though not from the same primary sources. Rather, all Suffolk County's water supply is from groundwater, and only one fourth of county residents have municipal wastewater systems, while the majority depend on about 400,000 aging septic tanks. Their human waste therefore contributes to runoff of fertilizer from both agriculture and domestic lawns, creating a complex picture of pollution. Multiple independent sources of nitrate runoff from a widespread area then coalesce in a common underground aquifer, making the contamination difficult to track and challenging to address.

Overall, independent reports from nonprofit groups such as the Soil Association in the UK and the Environmental Working Group in the US have sounded the alarm that industrial farming methods are the worst cause of nitrogen overdose in water around the world.[10] These reports call for an overhaul of agricultural practices, limiting the use of synthetic nitrate fertilizers and adopting best safety practices, including organic farming methods. Excessive nitrates in drinking water present an unacceptable risk for waterways and fisheries as well as for human health.

A parallel between the river systems of the earth and the urinary system of the human body offers revealing insight into downriver problems in both systems. In the human body, a substance that creates woe when oversupplied upstream is glucose. Like the gut, the kidney is designed to reabsorb sugars, but it can handle only a certain amount at a time. A deluge of sugars overloads the system, and they spill on through the body driving diarrhea from the gut or diuresis by the kidney. In both these cases, the sugar load also provides an unwanted loading of simple fodder for the organisms that live downstream in the colon or the urinary bladder, causing infections – colitis in the bowel and cystitis in the bladder.

The path to cystitis parallels the path to algal bloom in the Gulf. Too many simple sugars in the diet will be absorbed into the blood and spill over through the kidney into the urine, passing on to the bladder. Within hours, the sugars can function like gasoline on embers. Like dry wood on a forest floor, bacteria with the potential to cause an infection are always present in the bladder, but they are innocent and few, as is the normal amount of algae on a shoreline. But bring too much glucose into the human system and you light a fire, triggering an explosion of excessive bacterial growth and producing an outflow of inflammatory toxins with harmful effects – most typically, cystitis (urinary tract infection, such as we discussed in the context of antibiotics). For a person prone to cystitis or a doctor who diagnoses and treats it regularly, cause and effect become easy to spot: a rich meal on one day, perhaps with wine and two helpings of dessert, then symptoms of bladder infection the next. The runoff from Iowa takes longer to make its way to the Mississippi Delta than glucose from the human mouth to the urinary tract, but both follow a flow in one direction, with nowhere else to go as they approach the end, where they cause problems.

In keeping with the revelations of symmetry-seeking, the connection between the over-production of corn in Iowa and too much sugar in the human bloodstream is not just metaphorical but also literal. Corn, overproduced, has successfully sought a market and consequently found itself overrepresented in our diet. And as grains go, corn is disproportionately high in simple sugar; hence the development of corn syrup as a sweetener, then high-fructose corn syrup as an industrial-strength sweetener. Because corn is so cheap to produce, corn products have infiltrated the food market worldwide, not only as corn chips, popcorn, corn cereals and corn muffins, grits in the American South and polenta in Italy, and corn pretending to be a vegetable, but as a hidden ingredient in a wide range of foods, especially when transformed into a sweetener such as high-fructose corn syrup.

Corn as an ingredient is also tucked into salad dressing, gum, sweeteners of other kinds, toothpaste, dairy products, even whiskey (and corn has been put to countless non-food uses as well, most notably in vehicle fuels). But it is a high-calorie, low-nutrient food, so while it gives industrial farmers a great monetary return on their investment, it gives the consumer who eats all this corn a poor deal nutritionally. As we'll discuss in Chapter 8, relying too heavily on corn as a dietary staple can even result in serious diseases. But until we reverse the nitrate-driven overproduction of corn around the world, we will continue to see its problematic nutritional effects, ranging from that sugar-induced cystitis to the overall rise in obesity.

Pouring the synthetic chemical product of ammonium nitrate onto fields is now proven to feed this multi-tiered cycle of damage and unsustainability. The cycle starts with soil damage, then creates increased nitrate runoff in water supplies, with impacts on ecosystems all the way to the sea and the oceans, affecting both plant and animal life, including human life. Not only local and regional, synthetic nitrification is a problem of global proportions. From the loss of fish and other marine life to the contamination of drinking water and the invisible omnipresence of corn in our food supply, the direct and indirect effects of too much nitrate on human health are immense.

Hormone chemicals for bugs: DDT

Building on early experience with synthetic nitrate, DDT was another human-made chemical designed to fool mother nature and solve a knotty farming challenge. Also hailed as a scientific wonder of the mid-20th century, it was the first miraculous synthetic pesticide. Like synthetic ammonium nitrate, DDT (dichloro-diphenyl-trichloroethane) was first synthesized before World War Two, and scientist Paul Hermann Müller had discovered its potency as a pesticide while working for a Swiss chemical company. But its widespread use began in 1943, when the military adopted DDT to kill the fleas and lice that spread typhus and to wipe out mosquitos, which carried yellow fever and malaria. It saved many military and civilian lives during World War Two, and, as with synthetic nitrates, DDT production ramped up during the war years, so both stockpiles and further production capacity stood ready at war's end.

Effective, cheap, and abundant, DDT was promoted as a simple solution to pest control problems large and small, at home and abroad, prompting enthusiasm and confidence. The relatively short timeframe of the war had revealed little of its toxicity to humans, so the benefits of DDT seemed clearly to outweigh any theoretical concerns about possible long-term consequences. The general postwar excitement for US scientific innovations also opened the door wide for marketing and commerce to overshadow any apprehensions held by scientific researchers. Its feats as a fast-acting pesticide led to its promotion and popularity for all types of insect control – in homes and gardens, towns and beaches, and of course crop fields and livestock businesses.

Researchers issued their first warnings about DDT in 1950,[11] but its use continued apace. Then in 1962, biologist Rachel Carson sparked widespread public concern over the dangers of synthetic pesticide use and the urgent need for better controls, bringing the dangerous effects of insecticide abuse to public attention in *Silent Spring*. The spring of the title had become a silent season because a poisoned habitat meant not only poisoned insects but also dead birds and sick mammals. Carson's groundbreaking book spoke to how we are all in this together. Such concerns led to congressional hearings and ultimately triggered the establishment of the Environmental Protection Agency (EPA). Ten years after *Silent Spring* and considerable scientific reevaluation, the EPA banned DDT for agricultural use in the US in 1972, citing adverse effects on wildlife as well as human health risks.

Today, 45 years after the ban, we still see the long-lasting residue of DDT's relatively brief time in use. A breakdown product of DDT, called DDE, still shows up in 60% of samples of heavy cream from US dairy cows, 42% of kale greens, and 28% of carrots. According to the CDC, 99% of human blood samples in the US test positive for DDE.[12,13]

Women who were exposed to DDT as young girls are five times more likely to develop breast cancer, and it is also implicated in problems of miscarriages and low birth weight, male infertility, nervous system damage, and developmental delay in children. DDT has a chemical structure very similar to the hormone estrogen, and when given to young cockerels in 1950, it was shown to upset normal sexual development; they failed to develop normal testicles or the usual male characteristics like a colorful neck wattle, and the birds became sterile.

Investigating DDT's impact on human infertility, scientists have also evaluated the quality and quantity of sperm, analyzing semen studies from 1938 to 1991.

Not only were there losses of semen quality and quantity, an overall decrease in fertility, a rise of testicular cancer in men, and an increase in children's genital abnormalities (such as undescended testicles and hypospadias), but they also found that the conditions in men mirror the early reports of cockerels treated with DDT.[14-16] This mirroring of abnormalities between cockerels and humans exposed to DDT points towards a possible explanation underlying the results of a recent meta-study revealing a precipitous drop in semen counts across Europe and the US since the 1940s, from 100 million per milliliter to 60 million.[17] As zoologist Louis J. Guillette put it in his presentation to a Congressional hearing, 'Every man in this room is half the man his grandfather was'. Such phenomena beg for broad, interdisciplinary, symmetry-seeking research to find causes and explore possible means of prevention.

As we saw in Chapter 4, hormone messaging is nature's way of communicating, making sure that the processes of tissue growth and repair, body plan assembly, and construction of the fetus are accomplished with all their steps and stages, in the right order and at the right time. If something disrupts or confuses the messaging, the products of construction – the body – will be disordered and incomplete. Medicine was slow to recognize the dangers of DDT and to recognize it as a potent hormone disruptor because it was not like an ordinary poison, and it did not hurt humans the same way it killed insects. But DDT's synthetic structure rendered it chemically stable and extraordinarily persistent, so this substance does not degrade and break down into harmless products; rather, DDT resists degradation and persists, and it is thus passed readily along the food chain. Not only does it poison the insect, but the same molecules are passed on to fish and birds that feed on insects and all kinds of wildlife, as well as the humans who in their turn feed on those animals.

The early success of DDT – a chemical championed as a simple, wondrous fix – blinded our eyes to the symmetry of the bigger picture: that in living systems, poisons are poisons. DDT was such a stable, persistent molecule that it would last for years, go anywhere it could go, get into anything it could get into, accumulate in our fatty tissue, and mess up our human cells and systems wherever possible. But we had failed to consider the possibility that something so toxic to bugs could be profoundly toxic to ourselves, and we therefore missed the wider ramifications for infertility and human development.

Not only did DDT affect unintended as well as intended targets, but the insecticide's success was also self-limiting, because insects were quick to retaliate with strategies of their own. As with penicillin, DDT's results during the early years were spectacular, before insects got to know their new adversary.[18] But just as bacteria learned to resist penicillin, requiring that we bring newer, more potent antibiotics to the fight, widespread application of DDT has likewise led to widespread resistance from many insect pest species, so that it is now a far less effective insecticide than before. We know because – in stark contradiction to the World Health Organization's worldwide ban on DDT for agricultural purposes, which finally came in 2001 – they still currently approve this chemical for domestic indoor spraying to kill bugs. This even

though DDT is known to be a potent, persistent pollutant that accumulates in fatty tissues and has profound, long-term deleterious effects on human reproduction, causes cancer in animals, and is classified as a probable human carcinogen by US and international authorities.

Hormone chemicals for weeds: 2,4-D

Still more valuable to farmers than pesticides are herbicides, because ever-present competition from weeds is an even greater challenge than insect pests that might chew and eat crops. Gardeners and farmers face a constant challenge of weeding, dealing with unwanted plants growing next to and in competition with the chosen ones. Unfortunately but unsurprisingly, the ammonium nitrate fertilizers in common use feed both the intended target crop and the unwanted bystanders. Somewhat ironically, then, the appearance of synthetic fertilizer prompted a new problem: heartier weeds. The war years had already begun developing an answer, though, for in jungle warfare, the military had needed exfoliating agents to clear ground cover in jungles, so they had already been conducting government-sponsored experiments during the 1940s to find appropriate chemical compounds for the job.

Wartime research once again translated into agriculture technology, this time with the development of '2,4-D' (2,4-dichlorophenoxyacetic acid). This first widely used chemical herbicide changed farming practice profoundly, by saving an enormous amount of labor. Now farmers could clear more weeds from their fields than any amount of tilling and cultivating could accomplish, with only chemical spraying.

As DDT worked for dealing with insects, so this herbicide – 2,4-D – also works by mimicking some of the actions of hormones. It stimulates uncontrolled cell growth that causes treated plants, on land or in water, to grow so fast that they can't support themselves and so die. Discovered in England in the 1940s and funded by government research, 2,4-D was not eligible for patent protection that might have limited its use, so any company is free to make and sell it, and they do: today it is an ingredient in more than 1,500 products, marketed around the world.[19] Its patented cousins – atrazine, introduced in the 1970s, and glyphosate, from the 1980s and better known under the brand name Roundup – have become market leaders and extraordinary money makers for their corporate patent holders. US agribusiness spends over $11 billion every year on chemical herbicides and pesticides.[20] Tasked with killing weeds but not crops, the new herbicide worked for only certain kinds of plants, sparing others, so it necessarily worked for only specific crops. It would kill off broadleaf plants such as soybeans as if they were weeds but left grasses like corn relatively untouched. Corn – that same plant that responds so miraculously to synthetic nitrate – therefore gained another vote as the industrial farmer's crop of choice.

Roundup and GMOs

Ignore that poisons will be poisons, see the promise of enormous profits to be made by expanding industrial farming, and now imagine planning your agribusiness strategy. In a world unfettered by nature's obligatory symmetries and counterbalances, you would want to design plants armed with a built-in antidote, which carried a shield within them to protect against your own poisons. Enter genetic engineering and genetically modified Roundup Ready crops, introduced in 1996 for soybeans and in 1998 for corn.[21] The plant crops are armed with new genes to resist the action of Roundup, so that feeding with fertilizers for faster growth and spraying herbicides to wipe out weeds can happen simultaneously. What could be wrong with that picture? The farmer improves crop yields but with much less work, as feeding and weeding can be accomplished in the same spraying, and the farm can deliver more food for less cost.[22]

A habit of symmetry-seeking – of staying alert to patterns reiterated across species and to downstream consequences – points to an answer. Like

every organism, weeds will find their own adaptive strategies for dealing with a new problem, and, in a pattern that will be familiar to you now, developing resistance to the new synthetic threats is a usual strategy. So today we see Roundup-resistant super-weeds already, and farmers are back to tilling and to spraying herbicides that are even more toxic, in an effort to control the herbicide-resistant weeds. Some of these super-weeds can be killed by higher doses of glyphosate, so some farmers are spraying with even more Roundup than before, as well as using other, more potent chemical herbicides to try to cope with the problem of resistance.

Notably, when we consider the farmer's costs and returns, Roundup Ready corn is not more productive or more vigorous than any other corn, except in its resistance to this manufactured synthetic weed killer, Roundup. In fact, this synthetic hybrid species is less productive in one crucial way: its hybrid seed is somewhat impotent, because it does not produce uniform seed for planting. Only about one half of the next generation of plants will have the needed hybrid genome and expected phenotypic qualities and features for which the corn was designed. Conveniently for the single corporate owner who holds a monopoly on both patents, farmers must therefore purchase new seed corn every year, rather than following the age-old farming practice of keeping some seed from the best plants of this year's crop for next year, thus driving up their costs in a new way. The appeal of farming with chemical herbicides is that it is much less labor intensive than plowing and handpicking weeds out of the soil. But this short-term strategy of a shortcut to greater profits with much less effort is now revealing its unintended and unwelcome consequences and long-term costs, not only to farmers but to the environment and, inevitably, to human health.

Roundup Ready crops are responsible for the Roundup-intensive weed management practices that have accompanied them, and the US Department of Agriculture has estimated that farmers have used 380 million more pounds of herbicides than if

Roundup Ready crops had never been introduced.[23] As with nitrates, this increased use of glyphosate then produces more chemical runoff into creeks, rivers, and ecosystems. And whatever early environmental benefits Roundup seemed to promise, in addition to short-term cost savings on labor – of reduced tilling and less use of even-more-toxic herbicides – are fading, because the very weeds that Roundup was supposed to control have already sprung up in revolt.

Towards sustainable practices

Agriculture must be efficient and highly productive to produce food for the world and make a living for farmers, but to sustain productivity for more than a few decades, agriculture must also protect our most basic natural resources, particularly the soil and water quality on which future productivity depends. Travel through Iowa, Indiana, Illinois, Missouri and Nebraska and see hundreds of miles of American corn belt, fields after endless fields of this one crop. The midwestern US has grown corn since 1850, but only recently have only a few varieties – the GMO hybrids – come to dominate the landscape with uniformity.

Sustainable farming practices mirror nature's diversity. Hybrid monoculture protected by multiple synthetic concoctions may work well for a time, but predators and pathogens lurk in the wings to challenge, compete with, and threaten the arrogance of our human endeavors. Sooner or later, drought, disease, or other adversity will strike. When it does, all plants in the uniform monoculture will be equally vulnerable. Whichever species once stood together will fall together, for they share the same contiguous fields, identical genetic makeup, and physical characteristics in their densely packed ranks. Their uniformity renders them all equally susceptible to death or disease.

Global population growth drives demand for more food, so we continue to need increased productivity, but we must seek innovations that simultaneously create sustainable methods of crop production. To date, selective breeding of crop species and genetic modifications, enhanced soil fertility with synthetic fertilizers and irrigation, and weed and pest controls with chemical products have been the hallmarks of modern agriculture. With these methods, world food production has almost doubled in the past 40 years, but diverse ecosystems have been replaced in many regions by oversimplified agro-ecosystems vulnerable to pest attack. In order to safeguard the high level of productivity necessary to meet the human demand, these crops require protection from pests and disease, and the risks are greater when crowded together in large-scale monocultures of genetically uniform plants, which are not serving us well – neither our health nor our land and water.

The farm, then, turns out to be a locus for the concerns raised about hormone disruption in Chapter 4; about antibiotic overuse and resistance covered in Chapter 5; and about the nitrates, pesticides, herbicides, and monoculture here in Chapter 6. And the farm is where we humans get our food. So Big Ag deserves our further attention. In Chapter 7, we'll bring the lens of symmetry still closer to humans, with a look at animal farming, which like plant farming has also changed radically since the end of World War Two. Taking the broad view, we'll also consider the impacts of those changes, especially on the rise of human diseases, and on medicine's muted response to the changing plight of our national health.

References

1. Bennett D (2013). West, Texas: The Town That Blew Up. Online: http://www.bloomberg.com/news/articles/2013-07-03/west-texas-the-town-that-blew-up.
2. Ortiz E (2016). Deadly West, Texas, Fertilizer Plant Explosion Was 'Criminal Act': Feds. Online: http://www.nbcnews.com/news/us-news/deadly-west-texas-fertilizer-plant-explosion-was-criminal-act-feds-n572231.
3. Hager T (2008). The Alchemy of Air: A Jewish Genius, A Doomed Tycoon, and the Scientific Discovery that Fed the World, but Fueled the Rise of Hitler. Harmony.
4. Amadeo K (2018). How the Dust Bowl Environmental Disaster Impacted the US Economy. Online: https://www.thebalance.com/what-was-the-dust-bowl-causes-and-effects-3305689.

5. Pollan M (2011). The Omnivore's Dilemma: The Search for a Perfect Meal in a Fast-Food World. Bloomsbury Paperbacks; p.45.

6. Foley, J (2013). It's Time to Rethink America's Corn System. Online: www.scientificamerican.com/article/time-to-rethink-corn.

7. Heisler J, Glibert P, Burkholder J, et al. (2008). Eutrophication and harmful algal blooms: A scientific consensus. *Harmful Algae*, 8: 3–13.

8. National Geographic Magazine (2017). Fertilized World. Online: ngm.nationalgeographic.com/2013/05/fertilized-world/Charles-text.

9. Drinking Water Blues: Nitrate Pollution from Coast to Coast. Online: http://www.gracelinks.org/789/drinking-water-blues-nitrate-pollution-from-coast-to-coast.

10. Soil Association. Food for Life. Online: www.foodforlife.org.uk/about-us/our-partners/soil-association.

11. Philp RB (2013). Ecosystems and Human Health: Toxicology and Environmental Hazards. Boca Raton: CRC Press; p.335.

12. Philp RB (2013). Ecosystems and Human Health: Toxicology and Environmental Hazards. Boca Raton: CRC Press; p.337.

13. Agency for Toxic Substances and Disease Registry (2002). Public Health Statement for DDT, DDE, and DDD. Online: https://www.atsdr.cdc.gov/PHS/PHS.asp?id=79&tid=20.

14. Skandhan KP, Sahab Khan P, Sumangala B (2011). DDT and Male Reproductive System. Research Journal of Environmental Toxicology, 5: 76–80.

15. Foster TS (1974). Physiological and biological effects of pesticide residues in poultry. In: Gunther FA, Gunther JD (Eds). Residue Reviews. Reviews of Environmental Contamination and Toxicology (Continuation of Residue Reviews), vol 51. New York: Springer.

16. Burlington H, Lindeman VF (1950). Effect of DDT on testes and secondary sex characters of white leghorn cockerels. *Proceedings of the Society for Experimental Biology and Medicine*, 74 (1): 48–51.

17. Salam M (2017). Sperm Count in Western Men Has Dropped Over 50 Percent Since 1973, Paper Finds. Online: https://www.nytimes.com/2017/08/16/health/male-sperm-count-problem.html.

18. Berry-Caban CS (2011). DDT and Silent Spring: Fifty years after. Online: jmvh.org/article/ddt-and-silent-spring-fifty-years-after/.

19. Cobb AH, Reade JPH (2010). Herbicides and Plant Physiology (2nd ed). Hoboken: Wiley Blackwell.

20. USDA ERS (2014). Pesticide Use in U.S. Agriculture: 21 Selected Crops, 1960–2008. Online: https://www.ers.usda.gov/publications/pub-details/?pubid=43855.

21. Union of Concerned Scientists (2012). Eight Ways Monsanto Fails at Sustainable Agriculture. Online: https://www.ucsusa.org/news/press_release/monsanto-fails-sustainable-ag-1368.html#.WqvRv0x2t9A.

22. BusinessWire (2015). Monsanto and Scotts Miracle-Gro Expand Long-Standing Partnership. Online: http://www.businesswire.com/news/home/20150520005658/en/Monsanto-Scotts-Miracle-Gro-Expand-Lond-Standing-Partnership.

23. Main D (2016). Glyphosate Now the Most-Used Agricultural Chemical Ever. Online: www.newsweek.com/glyphosate-now-most-used-agricultural-chemical-ever-422419.

Animal farming and disease

For eons, farming has worked with the natural cycles of earth. As vegetation dies and decomposes, it moves back into the soil to feed the myriad organisms and life forms there. Worms mix the layers, preparing the bed for new plants, cultivating soil texture, preventing disease, and releasing plenty of nutrients to support normal plant growth. Introduce into this balanced ecosystem an overload of ammonium nitrate, as we saw in Chapter 6, and myriad problems arise.

In addition to those complications that come with synthetic fertilizers – of plant overgrowth, soil depletion, resistant weeds, algal bloom, the monopoly on seeds, carbon emissions, contaminated water, and the need for ever-harsher herbicides – these chemical products also leave the ground more acidic, so they distort and disturb the diversity of microorganisms that live in soil. And though unique conditions in different geographical locations demand assorted soil types to meet diverse climatic needs, synthetic fertilizer is one-size-fits-all, oblivious to those requirements. In the loud public debate over synthetic fertilizers, the contrast between industrial monoculture and organic farming is stark, arguing for maximum possible yield today versus sustainable preservation of the soil for the harvests of tomorrow. Less talked about in the land use debate is the disconnect between the raising of plant crops and the raising of animals, onto entirely separate farms.

Severing the links between plants and animals

Historically, a farm not only rotated crops but also included livestock in its mixed land use, and their presence together was a lynchpin in its sustainable productivity. A traditional farm was a self-sustaining operation and therefore necessarily diverse; animals fed on plants, and plants were fed with fertilizer from animals. The farmer kept the two in balance. But an influx of cheap, over-produced, and over-promoted synthetic fertilizer has driven animals out, off the farm and into confined animal feed operations.

Symmetry teaches us that nature will work hard to correct asymmetries that throw systems off-balance, many of them complex interdependencies that developed over millennia. Inevitably, as we have seen in previous chapters, the short-term benefits of man-made change tend to create unintended long-term effects, with consequent challenges. And one of the major balancing acts that nature had been perfecting for eons was this symbiosis between plants and animals – a reciprocal relationship with which farmers formerly needed to cooperate, if they wanted their farms to be productive. But one of the effects of synthetic nitrate fertilizer has been to disrupt this link, by rendering animals seemingly unnecessary for growing plants. Therefore, in addition to promoting monoculture that grows the same crop in a given field, year after year, disrupting the interdependence between plant species, today's industrial practices also remove animals from the process of growing plants.

In the story of the missing links with animals, a bad actor reappears, one you will remember from industrial plant production in Chapter 6: genetically modified corn. Its spectacular response to ammonium nitrate unfetters land from animals' needs and unleashes monoculture on the soil, for growers willing to glue the dirt back together with synthetic fertilizer. With the land given over to this one crop, corn growers then look for markets in every sector of the food chain, and it has become not only the basis of countless human foods but also the staple diet for the animals we eat. And so this loop of monoculture feeds on itself, deepening the difficulties and exaggerating imbalances. Corn as feed in commercial beef, pork, and chicken production has led in turn to ever-more rapid animal growth, with record yields of meat and poultry, so our human fate is now tied directly into this dangerous growth spiral. We are all in this together, the animals and us.

On a traditional farm, without the ostensible luxury of synthetic fertilizer, animals raised alongside crops would feed on the parts of the vegetable plants inedible for humans, and their animal wastes would be used to enrich the pasture and the fields. This cycle was contained and sustaining: products of animal excretion and composts of stable wastes generated on the farm were added back to the land, enriching field soils, cultivating the loam, and improving the living biota, fertility, and drainage of the fields. Wastes did not accumulate or spill over, because they were recycled and consumed right there on the farm, and the cycle adapted to the specific geographic and climatic conditions of the farm's location.

By eliminating the need to return the nutrients in animal waste to the field to sustain the next harvest, widespread use of synthetic fertilizer has reduced the need for crop rotations and fallow fields. It is no longer inescapable that one must leave land unplanted to rest the soil for a while, if one hopes for a better future yield. Synthetic fertilizers have not only accelerated the pace of plant growth (like the hormones that have accelerated animal growth), but they have also broken a natural cycle between plant and animal farming, inviting their divergence instead. We are left with, on the one hand, industrial monoculture of highly specialized crops

such as corn and, on the other, completely separate and equally vast animal production industries for chickens, hogs, dairy and cattle.

For industrial crop farmers, this bifurcation brings economies of scale, with corn growers no longer needing to plow the land after harvest to turn the plant stem nitrogen back to the soil. Rather, they support their monocropping with regular additions of synthetic products, a practice which works well for a while. But over time, it will have negative impacts on soil quality, by reducing the soil's organic matter, increasing soil compaction, reducing its ability to hold water, and increasing risks of erosion. Like the human biome – the flora of living organisms in our gut, easily disturbed by monotonous diet or medications – the biota of the topsoil is made vulnerable by monoculture and chemical applications, leading to increased acidity and loss of micronutrients and minerals.

Taint of nitrate: feeding or poisoning?

Not only do plants need animal waste to grow strong, but animals need plants to use up that waste, to keep the animals themselves safe. Instead, we now have the reverse equation: rather than using up excess nitrate, current plant farming is producing excess nitrate, as we saw in Chapter 6. Like plants, animals need nitrates to make amino acids and build proteins and DNA, but too much nitrate can be toxic for animals. The added nitrate now widely used to propel plant growth inevitably carries through to farm animals and humans, both of whom rely on plants to provide their nitrates, and the results can be disastrous. Many plants are grown as crops, which will go on to be fed to animals or processed for human consumption. At extremely high levels, this same additive that turns foliage a rich green color can sicken animals and humans, turning skin around the lips and face dark bluegray and even causing death. The reason for this apparent asymmetry – why nitrate makes plants flourish wildly but is dangerous for animals – lies in nature's response to asymmetry and disproportion, demonstrated in this case by what happens to red blood cells, the body's transport systems for oxygen, when they face an unsustainable imbalance.

An animal's circulatory system is necessarily more complex than a plant's. Plants are made simply: the chlorophyll that colors them green is what allows them to trap energy from sunlight, which they use to convert simple nutrients, carbon dioxide, and water into complex sugars, carbohydrates, and proteins. They have no need for a heart, because they have no red blood cells to pump. Only fish, birds, and animals have a closed circulatory system, which we creatures need for carrying oxygen to the living tissues. Chemically, we cannot draw our energy directly from sunlight, so our breathing lungs draw oxygen from the air. The hemoglobin molecules in the blood carry the oxygen to the rest of the body, and they are what color the blood cells red. When, on its journey to all of the organs blood passes through the lungs, it becomes stuffed full of oxyhemoglobin, so the red blood cells become brightly colored. This blood on the way from the lungs is rich in oxygen to be distributed around the body. The blood returning to the heart is darker, having given up much of its oxygen to the tissues.

The excess nitrate in plants fed to animals becomes excess nitrate in animals, and they convert nitrate to nitrite, a substance that can interfere with the body's transport system, because nitrite blocks the ability of red blood cells to carry oxygen. In humans, one of the signs of nitrate toxicity is a color change: the blood turns from red to a chocolate shade, leaving the oxygen-deprived skin gray-blue, in a condition called cyanosis. The color change is most obvious around the lips, mouth, and mucous membranes, and infants are much more vulnerable to the toxic effects of nitrite than adults, particularly babies bottle-fed with milk reconstituted from nitrate-contaminated drinking water. Babies with nitrate poisoning not only have blue-gray skin but also develop difficulty breathing, and they may become irritable or lethargic. Depending on the severity of the poisoning and whether or not it is recognized quickly and treated appropriately, weakening can

progress rapidly to causing coma and death. Adults with health problems involving the lungs are also vulnerable to this threat.

Like infants, sheep and cattle can suffer the same fate if exposed to a diet too rich in nitrates or given water contaminated with excess nitrogenous wastes or fertilizer. Pasture treated with high concentrations of synthetic nitrate fertilizer must therefore be left for two to four weeks before animals can safely be allowed to graze or before grass cuttings from the fertilized fields are safe to be stored in silos. Weather conditions can determine how long the farmer needs to wait, because rain speeds plants' uptake of nitrate. During a drought, crops grow more slowly and contain a greater concentration of nitrate. Using fertilizer can add dramatically to the nitrate burden – in addition to animal waste, human waste, and industrial pollution.

Unfortunately, removing animals from the fields has severed the link to plants and put livestock out of sight, into confined feed barns, where their overcrowded living conditions expose them to a higher risk of illness and disease. Nitrite poisoning is only one concern among many, as these confined animal feed operations that raise livestock for meat and dairy are inhumane for both the animals and workers. The artificial division also lends itself to high risk of producing unsafe food products from these industrial-style operations.

Ag-gag regulations

While animal rights activists tend to garner the most press for exposing abysmal conditions in 'concentrated animal feeding operations', or CAFOs, the implications for food safety receive less attention. But the same conditions that make factory farms dangerous for animals breed disease for humans. Animals unable to stand – 'downers' – can be exhausted, sickened or infected with common ailments or rare disease such as 'mad cow disease' (bovine spongiform encephalopathy, or BSE). Held in confined cages, they often lie in feces when they cannot stand, and are more likely to become infected with bacteria that are dangerous to humans as well as to the fallen animals themselves. The USDA enacted rules banning the slaughter of downer cows and calves in 2009 and 2016, respectively. Downer pigs, to date, are not covered by the ban.

Evidence of unsanitary and otherwise illegal practices has come to light largely through the work of investigative reporters and animal rights activists working undercover, videotaping and photographing inside CAFOs and writing about abusive practices. Their work has spurred some of the legislative and executive action to ban worst practices.

But industrial agribusiness has pushed back hard. In addition to lobbying against safety regulations, they are lobbying for new laws that block anyone, including reporters, from seeing how they raise our food or even, in some cases, from voicing an opinion about their products. In a well-known case litigated on one of these statutes, Texas beef producers filed a lawsuit against Oprah Winfrey in 1998. They claimed that it had been illegal for her to state on her television show that she was never 'eating another hamburger', after hearing about cases of mad cow disease in England and the possibility of the same disease infecting the food supply of the US. The industrial cattlemen lost their suit. Another high-profile legal battle lasted from 2012 to 2017: producers of a cattle by-product (treated with ammonia and sold to school cafeterias and others as food) sued ABC for using a term coined by others to describe it, 'pink slime'.[1] Many saw this and the Oprah case as attacks on free speech. The ABC case ended in a settlement, after drawing considerable attention to what goes into a hamburger.[2]

Under pressure from Big Ag, multiple states have proposed or passed legislation to ban photography or other recording on or about CAFOs. Industrial farming calls these legal maneuvers 'farm protection laws', but others prefer a label coined by reporter and food writer Mark Bittman, 'ag-gag' laws.[3,4] They aim to criminalize undercover investigations into industrial animal feed operations, slaughterhouses, and meat-processing facilities across the US – or even bypassers' photography.[5] In 1990, Kansas was the first state to introduce a law making it illegal to enter a private animal facility and take pictures or video with, they said, an 'intent to damage the enterprise conducted at that facility'. CAFO owners know the losses they can incur if people know the truth about their practices: in that same year, a parallel case was filed in the plant farming world. Under one of these 'veggie libel laws', the apple growers of Washington state sued CBS for $250 million after a 60 Minutes episode alleged a risk of childhood cancer from the chemical spray Alar, used in orchards. The reporting made an impact: the case was dismissed only after Alar was taken off the market.

While legal barriers to photography are inappropriate in any commercial setting, denying transparency to consumers and hiding treatment of workers, they are particularly disturbing around farming and the food chain. Clearly the picture is not a pretty one, as confirmed by the few videos that have leaked out from undercover investigations.[6]

As more people begin to understand these risks to our food supply (also described in the chapters that look at hormones, antibiotics, nitrates, and other problematic chemicals), lawsuits filed to push back against barriers to transparency are beginning to have some success. In mid-2017, for instance, a federal court overturned an ag-gag law in Utah that sought to criminalize investigations into animal abuse in CAFOs, citing the law as a violation of the first amendment and as against the public interest in safe farms.[7] And other cases brought by Big Ag to restrict citizen knowledge of their production practices are failing to meet tests of constitutional law. In late 2017, a federal court ruled that a Wyoming statute limiting data collection on water quality and pollution unconstitutional in its violation of the first amendment.[8]

Behind closed doors

For industrial meat production, CAFOs house many thousands of cattle, hogs, or birds under intensive conditions of high-calorie, corn-based feeding, artificial lighting designed to promote appetite, antibiotics to head off disease, and growth hormones to pile on the pounds and shorten the interval to market. Concentrated animal feed operations bear a strong resemblance to the concentration camps of World War Two: bleak, inhumane, industrial facilities, hidden from public view.

As in wartime, dirty secrets demand close guarding, which explains why the conditions of these animal lots have moved behind locked doors and high walls. With modern agribusiness, food producers eagerly brought the war's industrial technologies into agriculture and with them the horrors of war onto the industrial farm. Our food sources are the prisoners of war, as feedlots for beef, hogs, and chicken have come to resemble the overcrowded concentration camps of wartime. The farms replicate the camps' cruelty, filth, and squalor, and what reports and videos have made it outside the farms show appalling abuse of the animals by the workers (and, in some cases, abuse of the workers). But in place of starvation we find forced overfeeding and the perpetual sunshine of artificial lighting, to stimulate the livestock's speedy growth.

Given the overproduction of corn worldwide, these factories of animal growth now rely on grain-based diets rather than on the grass pasture that cattle evolved to eat or the varied diets on which healthy pigs and chickens depend. The methods for delivering food to them are designed to drive ever-faster fattening, preparing the animals for earlier slaughter, with the growth-promoting hormones described in Chapter 4 also accelerating the path at every turn. And the whole edifice rests on the synthetic nitrates that continue to drive overproduction of the corn that is now their feed, in turn driving overgrowth of livestock and disturbing our human food chain.

Like dirty toilets, confined animal feed operations often stink of urine and feces, because the animal waste from these industrial-scale feed operations is far too copious to be manageable, and only a small fraction of it can be recycled as manure. The largest confined animal feed operations produce more sewage waste than a major city. The government

accountability office estimates that a feeding operation with 800,000 pigs produces more than 1.6 million tons of waste a year, significantly more than the annual volume of sewage waste from the entire city of Philadelphia's sanitation department.[9]

Animals leave behind a vast amount of waste, unlike plants, which are highly efficient at using sunlight to convert nutrients into new shoots and leaves, flowers, fruits and seeds that can be used as food. As any parent knows from diaper duty, only a tiny proportion of a baby's food seems put to use as nutrition; most seems to pass right on through. This inefficiency is typical of all animals, young or old, as our bodies attempt to convert nutrition into muscle, bone, sinews, or skin. And so most of what's fed to livestock goes to waste, and little is converted into edible meat or dairy for human consumption. Not more than 30% of nutrients fed to dairy cows are transferred to milk or dairy products. For meat production, the conversion rates are estimated to be about 20% for chicken, 10% for pork, and as little as 5% for beef.

In the US, livestock produce between 3 and 20 times more manure than people every year – as much as 1.2–1.37 billion tons of waste.[10] Sewage treatment plants are absolutely essential and obligatory for managing the water quality of cities, but no such treatment facility exists for treating the flood of animal livestock wastes and excreta. Most of the effluent is stored in temporary holding wells or ponds, and often the waste is chemically tainted by high concentrations of heavy metals and pharmaceuticals, including antibiotics. Like the toxic waste sites of industrial manufacturing plants, the industrial scaling and intensive methods of confined animal feeding have created a monumental challenge to those who come after and must try to clean up the mess.

Poisoning the water, polluting the air

The essence of the foul stench coming from waste in CAFOs is ammonia, which is excreted in urine and feces. And where there is excess ammonia, there is also excess nitrate, because bacteria act on the chemical components in ammonia, which releases nitrates from the wastes. Besides the nitrates from excreta, nitrate can also poison the feeding systems of operations with confined animals, so the nitrate content of feed and water must be carefully monitored to avoid inadvertent toxicity. Cattle are more vulnerable than hogs, because the bovine ruminant stomach contains many organisms that actively convert nitrate to nitrite in cows, so in severe cases of cattle poisoning, symptoms can lead to death in a matter of hours. The mechanism of action is the same as in humans, and the telltale signs are like those for a human infant with nitrate poisoning: blue-gray discoloration around the nose and mouth, breathing difficulty, and muscle twitching. A stricken cow will stagger and then collapse. Without urgent intervention, cattle poisoned with excess nitrate sicken quickly and die.

Nitrates are so effective at killing mammals that they are marketed as commercial poisons to farmers, who use them as lethal bait to control unwanted foxes and wild dogs. But the farms that raise our food are now bathed in this same ingredient, from synthetically fertilized corn and from animals' own waste.

As we saw with the nitrate problem in California and New York in Chapter 6, groundwater is a major source of drinking water in rural America. The EPA estimates that 53% of the US population relies on groundwater for drinking water[11] – too often polluted by synthetic fertilizers and fecal and urinary waste. CAFOs' contribution to the nitrate burden is thus a significant public health hazard, as groundwater can be contaminated by the runoff leaching from manure, as overflow through leaks or breaks in storage or containment ponds. In 2000, the EPA's National Water Quality Inventory found 29 states specifically identified animal feeding operations as contributing to poor water quality.[12] And in Idaho, a study of private water wells detected not only elevated levels of nitrates but also measurable levels of veterinary antibiotics.[13] In Des Moines, Iowa, the nitrate concentration in the drinking water has

risen so high that they have instituted exceptional methods of water treatment to try to reduce it to a non-toxic level, and the city has sued adjoining counties to limit their application of synthetic nitrate fertilizers, identified as a major contributing cause.

Nor is drinking water the only element that CAFOs contaminate: they also pollute the air in neighboring communities, with negative consequences for human health. One example is a measurable increase in the rate of childhood asthma reported by independent investigators.[14,15] Concentrated animal feed operations emit not only ammonia but other chemical pollutants, including particulates and dust, which cause problems for adults with respiratory problems such as chronic bronchitis. People exposed to particulate matter in inspired air over a long interval can develop reduced lung function, and smaller particles absorbed by the body can have an array of negative health effects, including chronic fatigue.[16] Current methods on industrial animal farms also contribute significantly to the load of methane in the atmosphere, a greenhouse gas released through both animals' gases of belch and bowel and from CAFOs' pits of manure. Such emissions pollute the air, contribute to ozone depletion, and aggravate global warming,[17] which the latest science confirms is a growing problem and a threat to human health.[18]

Antibiotics on the farm

Besides the threats of nitrite poisoning to animals housed in CAFOs, added nitrate contamination in water runoff from their manure, and pollution that causes health problems for humans that live near CAFOs, their overcrowded inmates also making one another sick with contagious diseases. Thus animals housed in these conditions require continual, preventive doses of antibiotics to control infection – including the healthy animals. Without giving these routine antimicrobials to the livestock, CAFOs are unsustainable, because disease would kill too many of them. In her history of meat production in the US, *In Meat We Trust*, Maureen Ogle discusses the use of antibiotics for growth promotion.[19] But, as we saw in Chapter 5, injudicious use of antibiotics leads to ever more antibiotic resistance, in organisms that then cause human illness, disease, and death. The growing prevalence of highly resistant superbugs makes a strong case for improved control and stewardship of antimicrobials, especially in farming.[20]

Antibiotic use in animal feed and antibiotic-resistant human infections are closely intertwined, as we saw in our close look at antibiotics earlier. Resistant strains of disease-causing bacteria can be transferred from animals to humans through their handling animals or carcasses or by their eating inadequately prepared meat or dairy products. This threat is important news for human health because fewer treatment options now exist for those infected with antibiotic-resistant organisms. Antibiotics administered to animals are often not fully metabolized, so the chemicals spill over to be excreted in their urine and manure. Animal wastes therefore pollute groundwater not only with nitrates but with antibiotics and antibiotic-resistant organisms, which leech into the water and pose a threat to humans.

Strangely, antibiotics serve another function in farm animals, beyond combating or preventing disease. For reasons not entirely understood, regular doses of antibiotics also promote growth – an effect which farmers prize highly, for financial reasons. Ever since scientists were surprised by this discovery not long after World War Two, antibiotics have been a standard component of animal feed. On the plus side, then, for industrial-scale, single-species operations for meat and dairy, profitability and productivity have increased dramatically. Since 1960, milk production has doubled, meat production has tripled, and egg production has quadrupled. It also takes much less time than formerly to raise a fully-grown animal, hastening and thus increasing the return on investment. In 1920, a chicken took 16 weeks to reach 2.2 pounds, whereas now they can reach 5 pounds in 7 weeks.[21] The returns have proven irresistible, notwithstanding the extreme dangers, which scientists have understood and

warned about since the beginning of antibiotics' infiltration into animal farms.

Because of concern for human health, there is a growing effort to eliminate the routine use of antibiotics in concentrated animal feed operations. Over 400 organizations have called for legislation to phase out non-therapeutic use of medically important antimicrobials in farm animals. They speak for precisely the interdisciplinary breadth of perspectives that symmetry asks us to seek, when we step back and consider a problem from multiple angles, with input from multiple fields of experts: well-respected, non-partisan organizations representing medical, public health, consumer, agricultural, environmental, humane, and other interests, including groups such as the Infectious Disease Society of America, the American Public Health Association, the National Association of County and City Health Officials, and the National Sustainable Agriculture Coalition. In 2001, the American Medical Association approved a resolution to ban all low-level antibiotic use in animal feed, and the USDA has developed guidelines to help. The World Health Organization declared its opposition to the use of antibiotics in animal feed in 2003.

The dangers of diseases bred on our industrial farms are greatest to consumers – to the whole population – and when disease lands on humans, they land in turn in doctors' offices and hospitals. With such severe consequences for human health arising from current industrial farming practices, we would do well to ask why the doctor's voice is not louder in the fray of protest for prevention – for change. To take only one of the devastating downstream effects of the modern industrial animal farm, the problem of antimicrobial resistance is so serious that it threatens the very practice of modern medicine, as confirmed by the World Health Organization in their 2014 report.[22] Without effective antibiotics, more and more common infections could become untreatable or even fatal, and 'medical advances such as joint replacements, Cesarean sections, organ transplants, and chemotherapy could become nonviable'. Already in my medical practice

I see procedures that were once routine, such as biopsy of the prostate, now managed as potentially life-threatening.

But while doctors and medical researchers have finally begun to add their voices to the public debate to end unnecessary antibiotic use, as in the AMA's statement on antibiotic use for farm animals, they remain slow to take a firm stand on hormone disruptors such as herbicides and pesticides or plastics. And too many physicians remain silent, like bystanders looking on, reluctant to engage their patients in conversations about diet, lifestyle, and overall health.

Meanwhile, the amount of antibiotics manufactured and administered, without appropriate caution for the dangers already manifest, remains staggering. In 2011, the Food and Drug Administration reported that '13.5 million kilograms of antibacterial drugs were sold for use on animals raised for food in the United States in 2010' and '3.3 million kilograms of antibacterial drugs were used for human health in 2010'.[23,24] By their calculations, more than 70 percent of the antibacterial drugs sold and distributed in the US in 2010 were used on food animals, and only about 20 percent for human health. And, as we saw in Chapter 5, unless our decision-makers can begin to control the indiscriminate use of antibiotics, we will deepen the risk of a new era of plague and pestilence.

US legislators have repeatedly proposed a ban on the routine administration of antibiotics to livestock. Since 1999, Congress has repeatedly been presented with a bill to limit antibiotic use to only sick animals,[25] and it has the support of over 350 health-related and environmental groups. The legislator who is lead sponsor of the bill was a microbiologist, and the current draft seeking to make its way through the congressional voting process includes potent evidence, supported by the best science. It includes data from more than 40 years of peer-reviewed research showing that antibiotic use in animals is contributing to antibiotic resistance in humans. And that resistance is having an impact: 'In 2013, the Centers for Disease Control

and Prevention estimated that antibiotic-resistant infections cause at least 2 million infections, 23,000 deaths, 8 million additional hospital days, and $20 to $35 billion in excess direct health care costs each year in the United States'.[26]

If passed, the act would ban seven classes of antibiotics from being used in animals and would restrict the use of other antibiotics to specific therapeutic and limited preventive uses. Reintroduced every year, including this year, it still has not been voted into law, because multiple moneyed interests, including Big Ag and Big Pharma, lobby against it. More than 40 years ago, the FDA had already concluded that 'feeding livestock low doses of antibiotics used in human disease treatment could promote the development of antibiotic resistance in bacteria', and the death toll of US citizens is mounting, but Congress has enacted no significant legislation to stem the tide of damage. Meanwhile, more Americans continue to die from antibiotic-resistant bacterial infections every year than from HIV and AIDS.

The European Union has been more prompt to act on these threats than the US, and in 2006 they banned the routine use of antibiotics for growth promotion in agriculture. Since the ban, antibiotic usage has decreased, and livestock production has maintained its previous levels. And the statute is stemming the tide of dangerous infections: in 2010, the Danish Veterinary and Food Administration testified that the Danish ban of the nontherapeutic use of antibiotics in food-animal production had resulted in a marked reduction in antimicrobial resistance in multiple bacterial species, including *Campylobacter* and enterococci, both of which can cause dangerous food poisoning.

In response to public pressure, some major meat buyers (including McDonald's) have changed their policy and stopped using meat that was given the kinds of antibiotics also used for treating human infections. And some major livestock supplies are changing their ways too. Meat labelled as free from antibiotics is slowly becoming easier to find in some grocery stores, and after years of promising to reduce antibiotic use, some producers are finally making good on the pledge, with guarantees that they will not use routine or preventive antibiotics in their chickens, with Perdue advertising 'no antibiotics ever' and Tyson eliminating at least those antibiotics also used in humans, so far.[27]

Unfortunately, the political influence of the industrial agriculture corporations yields such might in Washington that Congress is still waiting to enact legal protection for US citizens. An annual FDA report on antimicrobials used in 'food-producing animals' reveals that instead of moderating downwards, the use of medically important antibiotics in food-producing animals in the US actually increased by 26% from 2009 to 2015. In that year, approximately 21.4 million pounds of these drugs critical to human use were administered to food animals.[28,29]

As with other problematic food production practices, the reason that no progress has been made in reducing antibiotic use in agriculture – even though they cause a host of severe environmental and public health problems that are reaching epidemic proportions – is that moneymakers refuse to give up their unsustainable profits. They know that without antibiotics, their unsanitary CAFO system will collapse, back down to manageable-sized farms and more modest earnings. And with their current outsize profits, they are able to buy the legislative influence to block regulation of food safety while enacting regulation against oversight or even reporting. Meanwhile, these overcrowded animal feed operations continue to make us increasingly vulnerable to deadly disease, particularly to infections.

Epidemics and pandemics

Bacterial infections are only one of the diseases that breed in tight living quarters, animal or human. Viral infections are another, and they have already shown their power to devastate overcrowded populations. One of the most familiar is influenza, or flu, a contagious respiratory illness that can be difficult to distinguish from other illnesses based only on symptoms. While some cases of flu are mild, those too can lead to a variety of complications, including other illnesses such as pneumonia, some of them serious or even deadly. Those with a weak immune system are more at risk, but a virulent strain can fell even the fit and healthy.

The worst influenza outbreak in recorded history, the Spanish flu of 1918, claimed many millions of lives during World War One, amidst the unique living conditions of the war. And it was human actions that invited the calamity, by bringing soldiers into proximity with the threat. Prior to infecting humans, Spanish flu was exclusively a swine flu, affecting only hogs, and in fact only one species had contracted the virus. But governments failed to foresee the likelihood that a swine flu could cross over to humans, with our mammalian physiology that shares many symmetries with hogs. During the war, pig farms were brought in too close to the battle lines, to provide fresh meat for the troops. Once unleashed in one area, this novel germ could spread too easily: with no prior exposure to this unfamiliar virus, the human body had developed no effective immunity to it – no circulating antibodies to provide a natural defense for resisting infection. Troops and refugees were packed into camps and on ships by the thousands, and millions of people were in transit, able to carry the disease far from its origins on the front. With these perfect conditions for spread, more lives were lost to Spanish flu during and after the war than to fighting.[30]

As biologists have come to understand the nature of viral disease, it has become clear that the origins of Spanish flu were not unusual; many human diseases are spread from animal reservoirs to humans. Virus particles don't care who the host is. They are opportunists, unable to reproduce on their own, but once taken into living cells they can redirect the host cells to make thousands of new virus particles, filling the cells which swell and burst open, releasing millions more virus particles to infect the next.

Wild animal populations demonstrate the destructive power of sporadic viral infections, which commonly fall prey to epidemics. In rabbits, a poxvirus called myxomatosis is as dangerous as influenza is to humans, so when Australia was overrun with rabbits in 1950, the government intentionally introduced the virus in an effort to control the mushrooming hordes of wild bunnies. Mosquitos did their part, spreading the infection, and this targeted biological war on the rabbits was devastatingly effective: the population was decimated, killing off 500 million in the first two years alone. Such is the power of novel viral infections. But if you are honing your habit of symmetry-seeking, you may guess what has happened since new, resistant strains have developed, and researchers have looked for – and found – ever-more deadly biological agents to try to kill the pestilent rabbits, without adequate consideration of the whole ecosystem within which the bunnies and the viruses wreak their havoc.[31]

An outbreak becomes an epidemic when it spreads beyond a small local area and overtakes a wider geographical region, affecting populations across large cities and whole states. And some epidemics reach beyond the regional to become pandemics. Like the Spanish flu of World War One, a pandemic spreads across countries and then continents to establish a worldwide footprint, presenting a global threat. In an era of unprecedented international travel and agricultural exports, epidemics are now much more likely to threaten or achieve pandemic status than at any other time in earth's history.

In recent years, H1N1 swine flu and H5N1 bird flu have proven the most destructive viral diseases. Patterns of spread are typically from species to species, and a common route is from chickens to hogs then to humans. Medical science works hard to try to monitor viral infections across the globe, to characterize and predict the patterns of viral cultures, to grow the virus particles, and to attenuate their virulence. The goal is to vaccinate vulnerable human populations against the likely strains, ahead of the annual flu season. One hundred years after the worldwide spread of the Spanish flu, vaccination is still the mainstay of modern medicine's strategy against another devastating flu pandemic. The most dangerous viruses will be those for which there is no vaccination, and we have only a small array of

antiviral medications that can successfully treat a virus after it has been contracted, so, when we have them, our best personal and national defense against the flu is to bolster our immune defenses with vaccinations.

Another of our best strategies for preventing epidemics and pandemics is to reduce the likelihood of cross-species contagion, but vast industrial animal feed operations have been carrying us far in the opposite direction, exposing us to contact with pathogenic organisms in meat and animal products, while actively cultivating antibiotic-resistant strains.

Germs love monoculture and crowding

The first virus ever identified appeared on a farm. Despite the development of some vaccinations for viruses by the end of the 1700s, it was not until 1892 that the cause of diseases such as smallpox, rabies, and polio was isolated and given the name 'virus'. This first recognized virus particle was called 'tobacco mosaic' virus, because of the mosaic pattern it marked on growing tobacco leaves. For a tobacco farmer, the leaf is the crop – the saleable part of the plant – so this infection was costly. As with viruses in animals and humans, its infection particles pass from the sick plants to the healthy, whether through contact with another plant, through insects, or through the farmer's hands that touch the plants, so it spreads easily when the plants are gathered closely together, as they are for growing. Like troops in a barracks or pupils in the close quarters of a school, where sneezes can readily pass infection from one to the next, a cultivated crop gathers equally vulnerable plants all together in rows. The farmer sows fields of like species and can only pray that viral infection will spare his crop.

Even corn is not immune from viral infections. Two viruses of special concern for corn growers are sugarcane mosaic virus and maize chlorotic mottle virus, and together they cause a disease known as maize lethal necrosis. It is hard to overstate the devastating impact of maize lethal necrosis in East Africa, where corn is a staple food and essential for food security. An estimated 19.5 million people in the arid lands of Uganda, Somalia, and South Sudan have suffered extended drought in the past decade, and this disease has added to near-total losses of the cornfields in some districts. Since its first appearance in Africa in 2011, maize lethal necrosis disease has already begun to spread, contributing to economic and political unrest as well as population displacement. Strategies for controlling the virus are largely restricted to the use of insecticides to try to control the insect carriers that can spread the disease.

Industrial agribusiness thus creates the perfect setting for infection, whether bacterial or viral, and not only for plants. Tens of thousands of the same species of animal crowd together in these densely packed feedlots or massive animal barn complexes, proverbial sitting ducks for infection to strike, contracted from a passing wild animal. In April 2015, the USDA reported an outbreak of bird flu in the turkey barns of Kandiyohi County in Minnesota, the largest turkey-producing state in the country. When a farmer first detected its presence, the disease had affected only one farm of 38,000 birds, but bird flu virus has multiple means of spreading: by contact between infected animals, in bird droppings, and through rats and mice. Within weeks, this highly contagious H5N2 strain had spread to 23 farms, causing a death toll of 1.2 million turkeys. Veterinarians scrambled to contain the outbreak by slaughtering all the birds on the affected farms, which stretched from Minnesota to the Dakotas, Missouri, Arkansas, and Kansas. Finally they contained the spread, but not before the bird flu had spread beyond the US to Canada. Minnesota is the land of 10,000 lakes, and scientists suspect that ducks or geese migrating across the water carried the epidemic over the lakes. In this outbreak, no human cases were detected in the US, though its patterns of spread across two countries made it difficult to track.

Where contemporary industrial farming methods invite epidemics of disease, more traditional methods fare better. One of the preventive treatments for tobacco mosaic virus is crop rotation, because nowhere in the natural world do you ever find a homogeneous expanse of a single variety of any living thing, since it represents an unsustainable challenge for the habitat. Nature abhors monoculture in both plants and animals. Why put all the eggs in one basket? Left to its own evolutionary devices, nature knows how to spread the risk and hedge its bets, ensuring that, come what may, some of each species will have a chance to survive. Monoculture appeals to our base human greed, like a gambler betting again and again on a combination that has been winning, and the US national agricultural policy of monocropping plays the same game – planting the same variety year upon year, raising more and more of the same animal species in ever-more crowded spaces – and thereby creates a potentially tragic vulnerability to disease, not only for the animals but for the humans who are vulnerable to the animals' diseases.

Monoculture invited the great potato famine in Ireland from 1845 to 1851, which killed over one million people and prompted the mass exodus of one million more. Potatoes were the staple crop in Ireland, and farmers favored a single variety, the Lumper, which had generated such a high yield for so many years that they had abandoned other types. But in the late summer of 1845, without any warning, the potato plants' leaves began to turn black and curly. The cause was wet rot, a fungal infection (*Phytophthora infestans*) – yet another category of disease, different from both bacterial and viral infections. Wind spread the spores and carried the fungus from diseased to healthy plants, and within weeks the nation's entire crop was lost. Unlike the prior sporadic crop losses that had affected only a few vulnerable varieties and only one district or another, wet rot had finally found its perfect target, and symmetry fed the flames. And because all the potatoes were the same – Lumpers – all were equally vulnerable to the fatal attack.

Unable to learn the glaring lesson from that tragedy, we nevertheless continue to plant all of one hearty variety of a crop, including potatoes.

For American farmers today, it's not the Lumper but the Russet Burbank – and not an accidental choice on their part but rather a prime example of the inter-systemic nature of our agricultural dilemmas: the Russet Burbank is McDonald's favored potato, so they purchase this preferred species to serve up 3.4 billion pounds of potatoes every year as French fries.[32] That decision, from such an economically powerful customer, drives monoculture, with all its attendant problems.[33]

All in this together

The dangers of epidemic infection are only one kind of the threat to human health that we face from CAFOs and industrial farms. If you step back to see the big-picture perspective, the disparate pieces of a vast and layered puzzle start to fit together, with a shared pattern of overfeeding, in system after system, and nature's attempts to correct the imbalance.

Knowing the architecture of humans and how we share most of our structure with other animals and many living things, you can understand what it means to say that we are all in this together, how hormones, antibiotics and other pharmaceuticals, synthetic fertilizer, pesticides and herbicides, and wastes on industrial animal farms all interweave to affect our food, our environment, and our day-to-day health. The vast scale of the challenge we must face in order to right the ship also comes into view from this distance, as we consider just how many industries must shift their practices towards health and sustainability. And industrial-scale agriculture emerges as central to the challenge, a focal point for needed change.

We have gained cheap and plentiful food from industrial-scale farms, and their corporate owners have reaped impressive profits, but this rapacious business model long ago passed the point of responsible citizenship. The flood of synthetic nitrates that allowed and propelled the split between plant and animal farming didn't simply drive small and medium-sized family-owned farms out of business: the new economies of scale, driven by massive corporate contracts, have driven the full chain of food production out of balance with the needs of all of us, the people who consume the food, the plants and animals that are the food, and the soil and water that produce the food.

Now chemical companies control the hybrid seed and dictate its GMO features that in turn promote sales of their corporate partners' fertilizers, herbicides, and insecticides. With the cooperation of governments, farmers, distributors, and consumers, corn – overfed by nitrates – has emerged as the totemic crop of monoculture worldwide, now dominating not only land use but the food supply of both livestock and humans. Animal production in large confined feedlots depends on overcrowding its inmates, leading to disease. Their vulnerability to infection inspires overuse of antibiotics, whose injudicious use is creating antibiotic resistance in organisms that cause human illness, disease, and death. Both CAFOs and monocropping lend themselves to the epidemic spread of viral, bacterial and fungal diseases.

And at the end of this long chain of excess we find today's biggest elephants in the room, the diseases of food. Heart disease, obesity, diabetes, and cancer are only some of the serious causes of disability and death that are clearly linked to what we eat. From the farm to our mouths, food will largely determine our life's health trajectory, and the science is absolutely clear: what you eat is the strongest predictor of your health and wellbeing. The drug-resistant infections now incubating in CAFOs are only one form of threat to human health that originates on industrial farms. The very foods that are the product of those farms are creating a new kind of epidemic.

Now that medicine has been rebranded as 'health care', with efforts at a focus on prevention, we might be tempted to turn over all thoughts about the health problems that begin on the farm to doctors and medical research. But to leave this issue to the medical industry is to fundamentally misunderstand what doctors are best at, where their priorities lie, and the nature of their expertise. In the next chapter, then, we'll consider what we know about medicine's responses to disease and epidemics.

References

1. Engber D (2012). The Sliming. Online: http://www.slate.com/articles/news_and_politics/food/2012/10/history_of_pink_slime_how_partially_defatted_chopped_beef_got_rebranded.html.

2. Kludt T (2017). ABC settles suit over what it had called 'pink slime'. Online: http://money.cnn.com/2017/06/28/media/abc-bpi-settlement/index.html.

3. Wilson L (2014). Ag-Gag Laws: A Shift in the Wrong Direction for Animal Welfare on Farms. Online: digitalcommons.law.ggu.edu/cgi/viewcontent.cgi?article=2126&context=ggulrev.

4. Bittman M (2011). Who Protects the Animals? Online: https://opinionator.blogs.nytimes.com/2011/04/26/who-protects-the-animals/.

5. Carlson C (2012). The Ag Gag Laws: Hiding Factory Farm Abuses From Public Scrutiny. Online: https://www.theatlantic.com/health/archive/2012/03/the-ag-gag-laws-hiding-factory-farm-abuses-from-public-scrutiny/254674/.

6. Greenwald G (2017). The FBI's Hunt for Two Missing Piglets Reveals the Federal Cover-Up of Barbaric Factory Farms. Online: https://theintercept.com/2017/10/05/factory-farms-fbi-missing-piglets-animal-rights-glenn-greenwald/.

7. Chappell B (2017). Judge Overturns Utah's 'Ag-Gag' Ban On Undercover Filming At Farms. Online: http://www.npr.org/sections/thetwo-way/2017/07/08/536186914/judge-overturns-utahs-ag-gag-ban-on-undercover-filming-at-farms.

8. The Natural Resources Defense Council (2017). Federal Court Rules Wyoming "AG-Gag" Law Unconstitutional. Online: https://www.nrdc.org/media/2017/170907.

9. United States Government Accountability Office (2008). Concentrated Animal Feeding Operations. EPA Needs More Information and a Clearly Defined Strategy to Protect Air and Water Quality from Pollutants of Concern. Online: https://www.gao.gov/new.items/d08944.pdf.

10. Environmental Protection Agency (2005). Detecting and mitigating the environmental impact of fecal pathogens originating from confined animal feeding operations: Review. Online: https://cfpub.epa.gov/si/si_public_record_report.cfm?dirEntryId=148645.

11. EPA (2004). Estimated Nitrate Concentrations in Groundwater Used for Drinking. Online: https://www.epa.gov/nutrient-policy-data/estimated-nitrate-concentrations-groundwater-used-drinking.

12. Congressional Research Service (2010). Animal waste and water quality: EPA regulation of concentrated animal feeding operations (CAFOs). Online: http://nationalaglawcenter.org/wp-content/uploads/assets/crs/RL31851.pdf.

13. Batt AL, Snow DD, Aga DS (2006). Occurrence of sulfonamide antimicrobials in private water wells in Washington County, Idaho, USA. *Chemosphere*, 64(11): 1963–1971.

14. Sigurdarson ST, Kline JN (2006). School proximity to concentrated animal feeding operations and prevalence of asthma in students. *Chest*, 129(6): 1486–1491.

15. Mirabelli MC, Wing S, Marshall SW, et al. (2006). Asthma Symptoms Among Adolescents Who Attend Public Schools That Are Located Near Confined Swine Feeding Operations. *Pediatrics*, 118(1): e66–e75. doi:10.1542/peds.2005-2812.

16. Michigan Department of Environmental Quality (2006). Toxic Steering Group (TSG).

17. Wolf J, Asrar G, West TO (2017). Revised methane emissions factors and spatially distributed annual carbon fluxes for global livestock. *Carbon Balance and Management*, 12: 16. Online: https://cbmjournal.springeropen.com/articles/10.1186/s13021-017-0084-y

18. U.S. Global Change Research Program (2017). Climate Science Special Report. Online: https://science2017.globalchange.gov/downloads/CSSR2017_FullReport.pdf.

19. Ogle M (2013). In Meat We Trust: An Unexpected History of Carnivore America. Houghton Mifflin Harcourt.

20. Khachatourians GG (1998). Agricultural use of antibiotics and the evolution and transfer of antibiotic-resistant bacteria. *Canadian Medical Association Journal*, 159: 1129–1136.

21. Pew Commission on Industrial Farm Animal Production (2009). Putting Meat on the Table: Industrial Farm Animal Production in America.

22. Prestinaci F, Pezzotti P, Pantosti A (2015). Antimicrobial resistance: a global multifaceted phenomenon. *Pathogens and Global Health*, 109(7): 309–318.

23. H.R. 1150 – 113th Congress (2013). Preservation of Antibiotics for Medical Treatment Act of 2013. Online: https://www.congress.gov/113/bills/hr1150/BILLS-113hr1150ih.pdf.

24. McKenna M (2011). News: FDA Won't Act Against Ag Antibiotic Use. Online: https://www.wired.com/2011/12/fda-ag-antibiotics/.

25. H.R. 1587 – 115th Congress (2017). Preservation of Antibiotics for Medical Treatment Act of 2017. Online: https://www.govtrack.us/congress/bills/115/hr1587.

26. H.R. 1587 – 115th Congress (2017). Preservation of Antibiotics for Medical Treatment Act of 2017. Online: https://www.congress.gov/115/bills/hr1587/BILLS-115hr1587ih.pdf.

27. Charles D (2016). Perdue Goes (Almost) Antibiotic-Free. Online: http://www.npr.org/sections/thesalt/2016/10/07/497033243/perdue-goes-almost-antibiotic-free.

28. Food and Drug Administration (2016). 2015 Summary Report on Antimicrobials Sold or Distributed for Use in Food-Producing Animals. Online: http://www.fda.gov/downloads/ForIndustry/UserFees/AnimalDrugUserFeeActADUFA/UCM534243.pdf.

29. The Pew Charitable Trusts (2017). Trends in U.S. Antibiotic Use. Online: http://www.pewtrusts.org/en/research-and-analysis/issue-briefs/2017/03/trends-in-us-antibiotic-use.

30. Wever PC, van Bergen L (2014). Death from 1918 pandemic influenza during the First World War: a perspective from personal and anecdotal evidence. *Influenza and Other Respiratory Viruses*, 8(5): 538–546.

31. Zukerman W (2009). Australia's battle with the bunny. Online: http://www.abc.net.au/science/articles/2009/04/08/2538860.htm.

32. What Variety Of Potatoes Did McDonald's Use? Online: https://idahopotato.com/dr-potato/what-variety-of-potatoes-did-mcdonalds-use.

33. PotatoPro (2016). The top six potato varieties grown in the United States (Fall Crop). Online: https://www.potatopro.com/news/2016/top-six-potato-varieties-grown-united-states-fall-crop.

Diseases of food: looking beyond medicine, to causes

Symmetry is about being similar and different at the same time. As we saw in Chapters 2 and 3, about symmetries in the body, human hands are this kind of similar: each one like the other, but opposed in a functional sense and pushing against each other, albeit often for common purpose. Such is a hallmark of symmetry, to be similar, similar, similar, and then different.

Health and wealth symmetry

Two of life's most sought-after states, health and wealth, follow this same pattern of symmetry between them. For young adults, being well is often the ordinary condition of living, one which they can take for granted. But ageing brings less functionality and more infirmity, and health then asserts its true value: priceless. Health and wealth are both highly prized, though in a head-to-head contest health always wins out over wealth as the more basic foundation of happiness – Virgil tells us that the greatest wealth is health. But when we plan our personal life objectives, defining financial goals is common, while designing a plan for lifelong wellness and health is rare.

The fundamental factors that shape wealth and health are similar. A good start in life is a boon to either: initial conditions strongly predict future performance, setting the trajectory. The fortunate start with some assets, to be nurtured, protected, and taught how to build good habits beginning early in life, knowing that today's saving is tomorrow's reward. Parental circumstances offer advantage or disadvantage, predicting likely outcomes. Both favorable genetic characteristics and family wealth can be passed from one generation to the next, helping to promote a secure future. But likelihoods are not certainties: a gambler can squander an inheritance just as substance abuse can poison a healthy body, and calamity can befall even the best-prepared, as external and unexpected conditions enter the picture too.

For those fortunate enough to have a little wealth, banks and modern investments can store our assets off-site and spread our risk. But not so for health, which doesn't allow us to save, store, or stockpile it. With the body as our only home, factory, and warehouse all rolled into one, we must store our energy right on the premises and carry it about with us, wherever we go. Too many reserves can burden the frame and overload the suspension, and overabundant stores will spill over and gum up the inner workings. Balance is the key for health and wellbeing, especially the balance between intake and expenditure: neither too much nor too little food or exercise. And while we know our banker cannot hold our hand when we walk through a store making spending decisions, we may lapse into assuming that our doctor is responsible for managing our health, forgetting that the absence of disease is primarily a function of a dozen small choices we make for ourselves every day.

After all these symmetries and parallels, health and wealth finally hit asymmetries, at points where they intersect, one directly affecting the other. Two effects are most notable, as well as contradictory. One is the impact of wealth on health, which is usually positive: around the globe, an overall rise in one's standard of living and a general improvement in a community's quality of human life both contribute to greater health. Widespread increases in wealth allow for better nutrition, better hygiene, better health education and information, and better medical care – including preventive measures – all helping to insulate us from certain diseases. But, somewhat paradoxically, affluence then carries its own health risks, so the rich have a long history of painful diseases. Gout is one, long known; too much wine and rich food (such as were once available to only the well-off) can increase uric acid retention, depositing sharp crystals in the joint space at the base of the great toe, where they cause an intense reaction of bright redness, swelling, and excruciating pain, or triggering sudden colicky pains in the flank, as stones form in the kidneys and obstruct the flow of urine. This classic affliction of bishops and kings is a disease of excess.

Diseases of poverty, diseases of affluence

In impoverished nations, on the other hand, disease is an immense challenge, with multiple deadly diseases arising from global poverty today, the most common of them AIDS, malaria, and tuberculosis, together accounting for 1 in every 10 deaths worldwide each year. Measles, whooping cough, and polio also still exact a heavy toll, though there are well-known, effective treatments and prevention for these childhood diseases, making them the first targets for intervention from global health initiatives, which have taken great strides towards reducing their incidence.[1] Nutrition is vital for successful outcomes, because while well-nourished children may become ill from measles they rarely die of it, since resistance arises to fight off the effects of a pathogen if its human host is healthy. Just as the mobile home park sustains more damage from a tornado than do

houses of stronger construction and the people on the flood plains of Bangladesh suffer greatest losses in the floods of the rainy season, initial conditions set the scene and will largely determine the impact and severity of disaster or disease.

Diseases of wealth, on the other hand, differ from diseases of poverty, and as affluence rises around the world, so do its diseases. Where once we had a handful of health woes linked to wealth, we now have a panoply. Common illnesses of affluence include obesity, hypertension, heart disease, stroke, dementia, and cancer, with multiple factors conspiring to promote these conditions. Many are problems of diet, from foods now known to carry greater risk of disease.

Food marketers, in promoting their wares, eagerly cast doubt on what we know, but researchers without vested interests repeatedly report the same general themes about what foods are dangerous.[2] Even amidst the dueling camps of dietary fads, considerable research now confirms that foods with more fat and higher sugar content, consumed too regularly over time, carry threats to health. Fare heavy in refined white flour, calorie-rich offerings such as meat and dairy, and processed foods deliver a higher risk and offer a greater burden of disease. And onto the ostensible luxury of a sedentary lifestyle, we also replace physical exertion with increased use of alcohol and tobacco products, both known to have significant negative impacts on health.[3]

In America, the modern epidemic is obesity. More than one-third of US adults are now obese, and the percentage of children with obesity in the United States has more than tripled since the 1970s,[4] so that today, about one in five school-aged children (ages 6–19) has obesity.[5] To call it a disease of affluence hides the reality of its toll, however, because while access to problem foods makes everyone in a wealthy country vulnerable, the poor are at greater risk than the well-to-do. The two US states with the highest rates of childhood obesity, Mississippi and Georgia, are not two of the nation's poorest by coincidence.[6-9] Obesity and malnutrition develop hand in hand when the body does not receive the nutrients it needs in the amounts it requires for healthy tissues and full function; too much is just as bad as too little. Good nutrition is the foundation of health, but it is a casualty of poverty in a wealthy country. And illness itself, like poverty, can trap you and limit your freedoms of choice and opportunity.

Even before obesity was officially labeled a disease in 2013,[10] overweight and corpulence have long been known to link closely with other conditions that lead to loss of life and lost quality of life: hypertension, heart disease, cancer, diabetes, and numerous other diseases are associated with increased inflammatory markers and oxidative stress. The processed food that has proliferated in recent decades, simultaneously heavier with calories but poorer in nutrients than similar quantities of whole foods, is part of this landscape of obesity.

Since 1950, the cardiovascular problems that have arisen along with obesity in the US have made heart disease our most serious health problem, and in spite of considerable investment in research, treatment, and prevention, heart disease continues to be the number one cause of death in the US and the world.[11,12] The past 25 years have seen a somewhat better prognosis for those treated for symptoms of heart disease; a reduction in cigarette smoking, more emphasis on controlling blood pressure, and decreases in blood cholesterol levels (through both dietary change and medications that lower lipids) have reduced some of the risk. Any one cardiovascular event has also become less likely to kill you, as emergency medical care has improved to include diagnostic angiography, pharmaceutical interventions, and minimally invasive revascularization. But heart disease is still the most common cause of death for both men and women.

Food, the main contributor to obesity, is essential for the energy to stay alive, but it is not simply fuel: we rely on it to provide all of the essential building blocks for our bodies, including the many vitamins and minerals that the body itself cannot provide. For any life at all, we must rely on food to eat, water

to drink, and air to breathe, but for a healthy life, we need not just water but clean water, not just any air but fresh air – free of toxic chemicals, dangerous germs, heavy metals, and other pollutants – and the foods we eat must be nutritious.

Food as building blocks

To understand the necessity of a varied diet that includes a full mix of nutrients, we need only look to a disease of deficiency, such as scurvy. For centuries, when sailors embarked on months-long sea voyages, some of them would become weakened and ill with this disease. Its symptoms and signs include mouth sores and swollen gums, dermatitis, diarrhea, bruising and bleeding, joint pain, eye problems, and anemia, sometimes leading to death. And the history of scurvy's diagnosis, testing, and treatment proves an illuminating case study as we consider how to apply our principles of symmetry.

As we glimpsed during our dive into individual cell metabolism, in Chapter 3, the collective of cells that make up the adult body is constantly reshaped, its building blocks like Lego, to be taken apart and rebuilt again. Red blood cells, for instance, have a life of only about 120 days before they break down and have to be replaced. Diseases of nutritional deficiency thus routinely show signs and symptoms of anemia. Cells of the skin (the dermis) and the lining of the mouth are continually shed and replaced, so if some of their essential building blocks are missing, the architecture of skin and gums will change. Like the outer skin, the inner lining of the gut is constantly remodeled too, and without its essential elements, drawn from nutrients in the diet, the structure of new cells will be incomplete and lack its normal resilience, changing both the form and the functioning of the gut as well.

When we imagine an embryo or a growing child, it is easier to embrace the idea of building blocks creating a new human from scratch, with the nutrients from food as those multicolored blocks. But bodies, once built, also demand constant maintenance, renovation, and repair work. Like a multistory apartment complex or even an entire mixed-use city district, the body makes constant demands for new building materials, which can enter the premises in only one raw form: nutrition, in the foods we choose to eat.

Those sailors with scurvy were missing some essential building blocks: adequate amounts of vitamin C, found in most fresh fruits and vegetables, which were unavailable on their sea voyages. Humans need vitamin C for producing collagen – the molecular threads that the body weaves to make all kinds of connective tissues for fascia, tendons, teeth, and bone. Missing this vital nutrient, a person with scurvy first shows only mild symptoms that are subtle and hard to recognize, but over time the advanced disease becomes more florid and even life threatening.

In the modern world, patients with full-blown scurvy are rare, but as in all deficiency diseases, cases can range on a spectrum from mild to severe, and its milder features are not uncommon, even today. Investigators have recently demonstrated that pregnant women lacking vitamin C are prone to have babies with smaller heads and incomplete brain development, reflecting the pattern fundamental to symmetry-seeking: that deficits in earlier developmental stages will have more profound, sometimes devastating impacts. The insult of any dietary deficits will always land more heavily on the embryo, the fetus, and growing children than on adults, whose architecture is already fully formed – even if the result of those deficits in the fetus is not always apparent until later in the life cycle. Within the embryo, the brain is particularly sensitive to missing nutrients, and subtle features of dietary deficits are more likely to be manifest not only in brain size, as with inadequate vitamin C, but in cognitive dimensions of childhood development, such as attention deficit disorders or learning disabilities.

Food as toy building blocks is a useful metaphor, because construction may demand a specific size, shape, or color of brick, but little plastic bricks do not spontaneously click themselves together into structures, nor do the cells of the body, even when all the essential pieces are present. Each edifice also requires some kind of plan – orderly directions for assembly that must be communicated to the building site. In living tissues, DNA contains the blueprint. RNA conveys the messaging to drive the ribosomes' protein synthesis, but hormones control the timing, order of events, and communication of instructions on the work site, to guide orderly construction. Nutrients and calories from food are essential materials for bodily construction work, but nutrition is only one necessary component: the project will be in equal jeopardy if any one of several required elements is compromised – especially if hormone messaging is disrupted.

The slow pace of change

James Lind, a British Navy ship's surgeon, was interested in this disease of sailors, and he recognized that the longer their time at sea, the worse were their symptoms and the greater their risk of dying. Discerning a possible cause in their diets, Lind made himself one of the first in the history of medicine to conduct a proper 'controlled trial'. He divided sailors into equal groups, treating some and deliberately not treating others; one group he gave two oranges and a lemon to eat with their rations, the others some cider. After months at sea, recording careful observations, Lind's study showed conclusive results: sailors eating citrus fruits remained healthier, while many of the others developed signs of scurvy, and some of them had even died of it.

He published his 'Treatise on the Scurvy' in 1753, and you might reasonably assume that within a few months, the admiralty had instituted new rules regarding rations for sailors, but your confidence would be misplaced. Instead, it took almost 50 years before the British Navy began to adopt Lind's recommendation to provide citrus fruit as a standard ration at sea. Recognizing the cause and then finding, testing, and offering an effective solution proved not enough to prompt the change these sailors' health and lives demanded. Rather, the new ideas had to battle with then-prevailing scientific and medical beliefs and to change many minds, before they could change prevailing practice. Years had to pass to allow enough time for defeating the resistance of the senior practitioners with their fixed positions, giving doctors time to grow old and step down from their positions, retire, or die. Younger officers with fresh ideas, new knowledge, and open minds could then be promoted into positions of authority and effect change.

As we explore a few other well-known diseases in this chapter, you will see that this delay to adopt is typical, this reluctance to act on even well-tested evidence. Too often, the problems of a new disease, though recognized, have to get much worse before finally gaining the appropriate attention of the medical community, leading eventually to doctors at last embracing the more effective strategies. British admiralty leaders did finally take action and institute effective measures. Their choice of limes as the source of vitamin C – cheaper than other citrus, as they were more readily available from the West Indian colonies – led to the nickname of 'limeys' for Royal Navy sailors and, later, Brits in general.

Corn lacks vital bricks

The US medical literature first described a case of a new disease called pellagra in 1863, and by 1912 there were already more than 25,000 cases, with a mortality rate of 40%. Like scurvy, it was a gory disease, this one with severe intestinal symptoms, skin lesions, and mental illness, giving it the moniker 'the disease of three Ds': diarrhea, dermatitis, and dementia. When it first appeared, it was thought to be an infection, but then no doctors, nurses, or attendants ever fell ill. And instead of showing itself in densely crowded cities, where contagion would spread a typical infectious disease, pellagra occurred in more sparsely populated rural areas, and only poor people seemed to get it.

Pellagra also had a geography: it was a disease of the American South, with clusters of severe illness and deaths in government institutions such as orphanages, jails, and mental hospitals. According to estimates of the years between 1906 and 1940, more than 3 million Americans were afflicted with pellagra and more than 100,000 died.

In addition to this new human disease, dogs in households on farms with pellagra sufferers had a new disease of their own, 'black tongue', which left the animals lethargic, anemic, and wasted, suffering like the humans from diarrhea, as well as from sores about the mouth. For dogs, though, the most distinct feature of pellagra was a jet-black tongue. Like the human disease, it took its course over weeks or months, after which the dogs would usually weaken and die.

Campaigns to recruit soldiers for World Wars One and Two brought special attention to pellagra, because it rendered many young men too weak to serve in military service. In the search for a cause, it became clear that its sufferers relied heavily on corn in their staple diet of meat, corn meal and molasses, corn grits, and corn biscuits, with only smaller amounts of other foods, such as some cabbage or sweet potatoes. The afflicted dogs, too, shared the same corn-based diet. Those studying the disease, thinking of it as perhaps an atypical infection, sought some kind of poisoning from a toxin in corn that might be responsible. But the critical element turned out to be not something hiding in corn, but rather what was missing. Corn-based diets lacked an essential vitamin from the B family known as nicotinic acid or niacin ('NIcotinic ACid vitamIN'). As is typical in other nutritional deficiency diseases, the lack made an impact not in a single organ system but in many.

When patients were treated with niacin supplements and those at risk adopted a more mixed and nutritionally balanced diet, the epidemic finally resolved. It had taken 80 years since the disease had first appeared in the literature and 35 years after one Dr. Joseph Goldberger and his team had discovered how to cure and prevent it. The path to wellness was to correct for an asymmetry – the missing B vitamin. And just as deficiency can be devastating, causing disease and dysfunction as in pellagra or scurvy, excess can also be damaging, too. (Overdosing on a vitamin by eating vitamin-rich foods is almost impossible, but toxicity is common in those who take daily megadoses of vitamins over a long period of time.[13])

Infection as a cause of disease

Medical history offers tale after tale of the challenges of diseases: the difficulty of establishing that a condition is a disease, with a discoverable cause; the extraordinary amount of detective work that the disease can demand, to track down and identify its specific causes; and the struggle to convince contemporaries of its existence and best treatments – amidst resistance, institutional inertia, and the painfully slow pace of change.

Among the first kinds of disease to have a recognized cause were the sexually transmitted infections such as gonorrhea and syphilis. Signs might appear within a few days after the encounter between a previously healthy sexual partner and an infected person, manifesting as genital symptoms. This short latency between exposure and onset of symptoms – as well as the identifiable mode of transmission, through specific sexual contact, with symptoms in the area of contact – helped the patient to recognize a likely causal relation and pointed physicians towards a diagnosis. (An unwelcome intruder like any infection, venereal disease earned unique labels between nationalities, the English calling it 'the French disease' and the French 'the English', each blaming the other as its source.)

Subsequent opportunities to study infectious disease included epidemics, with germs causing not only illness but death, such as in the cholera outbreaks of London in the mid-19th century. Characterized by severe, colicky abdominal pains, fever, and bloody diarrhea, cholera infection often brought death within a few days or in some cases just hours.

Prevailing beliefs in those days held that disease was carried by miasma vapors or stench in the air, and, like all major European cities, London was a stinky place – reminiscent of the over-crowded animal farms of Chapter 7. As Steven Johnson depicts the city in *The Ghost Map: The Story of London's Most Terrifying Epidemic*,[14] it had no sanitary systems or waste disposal; privies emptied into shallow cesspits, open pipes or cellars, and garbage rotted in the streets. Diseases were most common in the smelliest parts of town, where the poorest neighborhoods stank still worse than others. Many assumed the poor to be morally depraved as well, a state said to make them more vulnerable to disease. Studying the outbreak of 1848 after it ended, however, Dr. John Snow developed a contrary theory: he suspected that drinking water might have been the cause.

In September 1854, another outbreak of cholera struck, this time in the more well-to-do Soho district, close to where Snow lived. This time, backing up from individual cases to see them in a larger context, he was able to draw clues from the geography of the disease. Cases clustered together in certain areas, where the cholera affected some but not all of the people who lived within a limited range. Mapping the area of central London and recording deaths as they occurred, Snow saw a pattern revealed: the cases appeared around Broad Street.

Within only a few weeks, there were 578 deaths. But at the epicenter of the outbreak, the brewery on Broad Street, not a single brewery worker had fallen victim to the disease. Apparently, they had some advantage over others in avoiding disease, which turned out to be the perk of free beer that brewers consumed, instead of contaminated water. A public water pump in the street outside the brewery was the source of the epidemic, having been contaminated by a burst drain on neighboring Albion Terrace that had bled sewage waste into the water supply, weeks before the infections appeared. Discerning from its patterns of infection that the disease was in the drinking water, Snow had the handle removed from the pump, and within days the epidemic was over.

After the crisis had ended, he published his street map showing the distribution of cholera deaths and presented the evidence that the disease had been carried in the water supply. He made the case for improving the quality of London's water supply and for better sewage disposal and waste management. His evidence that the contaminant lay in the water supply was more than strong – even though the specific causal agent was still unknown and unnamed, the case for its source was hard to dispute, when disabling the water pump had halted the epidemic.

But it was almost 50 years before the medical community would listen to Snow, and even longer before London's city officials would begin to embrace his theory and start to mobilize resources to engineer significant changes. They found the inconvenient news about risk in their water system annoying, as the proposed changes would be costly and disruptive, and anyway the epidemic had passed.

London did eventually reengineer their water and sewage systems, but elsewhere, in countries without clean water, cholera is still an important killer today. Now we know not only the delivery mechanism and how to stop an outbreak, but also the specific bacteria responsible for the disease (*Vibrio cholerae*), which Dr. Robert Koch isolated and confirmed in 1883, years after Snow's research. Nevertheless, the world still sees more than 4 million cases of cholera per year, and up to 142,000 deaths.[15]

Joining the dots in cancer cases

In the history of medical discovery, a telling test case is cancer, as it demonstrates the difficulty in recognizing symptom patterns that signal a disease, developing and propagating reliable treatments for it, and finally finding and promoting ways to prevent it. This dreaded disease – actually a collection of similar diseases, some with unique profiles – can occur in any part of the body, appearing as a wart, tumor or ulcer. Cancer may stay contained where it begins, or it may invade adjacent structures, traveling to other organs within the body, perhaps by spreading through the lymphatic system or bloodstream. Unlike infection, it does not pass from one individual to another, so it is not a danger to caregivers or other household members. Often, if a cancer problem presents and can be diagnosed, treatments

can contain and control the disease, even for some cancers that are difficult or impossible to cure. And research has finally taught us a great deal about prevention. In looking for the causes of cancers, a common trigger is often exposure to a particular chemical product or chronic irritant. Groups of workers can therefore be at increased risk for a particular kind of cancer, as they share repeated exposure to some trigger, such as asbestos, now widely known to be a cause mesothelioma.

In 1775, Sir Percival Pott described an unusual kind of cancer, unique both in its location – always on the underside of the scrotum – and because it occurred not in old men as would be more typical of cancer but in young boys. The disease appeared as a chronic warty growth that would ulcerate and bleed. Pott also noted his observation that all of the victims were 'climbing boys'. His was the era of *Oliver Twist*, and, like young Oliver, orphan boys were taken at a very early age to work with chimney sweeps, climbing up and down to sweep the filthy flues of London homes and smoke stacks.

Starting young and spending many years clambering inside chimneys, the climbing boys would make daily contact with soot that blackened their clothes and bodies from head to toe. It washed away more easily from the smooth skin of their faces, shoulders, and backs than from the deep folds of their hairy scrotal skin. And like smoke from cigarettes, smoke from coal fires contains not only soot and particles but also volatile hydrocarbons and tars, which provoke tumors and cancers when painted onto the skin.

This type of cancer in the climbing boys offered several unusual clues that were hard to miss: its unusual site on the body, the scrotum; the unusual age of those smitten, young boys; and the glaring coincidence that all the victims had worked for years in the sooty flues of London's dirty chimneys. These markers made it possible to join the dots and make the connection to its cause. Rather than scientific expertise, it was Pott's eye for pattern which noticed that it was young chimney sweeps who were at risk, just as it was not specialized knowledge that led

John Snow to the contaminated water pump but his eye for the geographical clustering of cholera deaths. We can all recognize patterns, whether in our own health or in public phenomena, and symmetry-seeking helps us to be even better at it.

Smoke screens and cancer

Pinning down the source of the climbing boys' cancer proved much easier than recognizing and responding to the link between cigarettes and cancer a couple of centuries later. The habit of smoking was spread wide, across millions of people. Many smokers never developed the diseases now linked closely to smoking, such as lung cancer, and some people who had never smoked did (and do) contract those diseases too. Today, most people are finally convinced that cigarettes are bad for you and that smoking can damage your health – that the behavior of cigarette smoking carries many health risks and increases the likelihood of many diseases, especially lung cancer, and of premature death. But even after the medical evidence was in, it took decades for doctors and researchers from other fields to persuade their colleagues of this fact – to establish the connection between tobacco and its risks so irrefutably and insistently that dangerous public policies and private behaviors began to change.

First reports linking cancer and tobacco products surfaced as early as 1928, on nasal cancer in snuff users. Then in 1938 we saw lip and tongue cancer in pipe smokers, and by 1948 the literature reported increased risk of premature death in cigarette smokers. By 1952, there was little doubt that this risk was real and significant. Still, stalwart resistance remained, and some voices vehemently denied the dangers of tobacco products. As with the causal connection between drinking water and cholera, science was very slow to arrive with the definitive proofs necessary to bring consensus across the medical community.

'Big Tobacco' fought for decades to redirect responsibility for the diseases of tobacco use, always blaming the smokers who chose to use their products,

but the sources of the problem were multisystemic. The US government promoted tobacco heavily by subsidizing farmers to grow it, and vast numbers of people therefore depended on tobacco products for their livelihood, involved as they were in the manufacture, promotion, and distribution of tobacco products. Starting in World War One, the military also allowed tobacco companies to target US troops with cheap and free cigarettes, even including cigarettes in soldiers' rations, alongside their food, in World War Two – a practice that did not flag until 1975, and which guaranteed that several generations of men and women would return from US wars addicted to a smoking habit.[16]

Commercial success gave the tobacco industry vast resources for political lobbying and influence, so the public received very mixed messages about the consequences of smoking – the louder voices more positive than negative. Some doctors aligned with Big Tobacco, and advisory groups of salaried spokespeople, ostensible experts, spoke out strongly in defense of smoking and tobacco products. Big Tobacco sponsored 'tobacco science' to publish favorable spin about the supposed health benefits of smoking, and others provided smoke screens and distractions to play down the associated health hazards, rebut concerns about addiction, and deny the risks of heart disease and cancer.

Massive advertising campaigns promoted tobacco products to young people and women, and reputable news sources supported the benefits of smoking. Promotional initiatives linked professional sports' celebrities to branded tobacco products with collectables such as cigarette cards. As news of any risk made it through the hype, tobacco stayed one step ahead, constantly presenting itself as a work in progress, ever improving, making newer and healthier products to try with lower nicotine content or lower tar, tipped with filters or flavored with menthol. Every new twist promised that less risk to human health was just around the corner, while Big Tobacco worked tirelessly to resist accurate labeling, disclosure of health risks, and explicit warnings on packaging. All the while tobacco continued to sicken and kill untold numbers of human beings worldwide.

In this environment of resistance to the truth about tobacco's health implications, the US government also proved reluctant to require that tobacco products be marked with warning labels of their dire risks, like those generally used elsewhere. In Britain and Europe, cigarette packets carry large-print cautions that 'SMOKING WILL KILL'. Unequivocal warnings take up at least one third of the package: 'WARNING: Cigarettes cause cancer' and 'SMOKING CAUSES FATAL LUNG CANCER', a skull-and-crossbones symbol, and the labeling which has proven most effective: graphic images of lung disease, from amputated toes to the postmortem face of a human corpse.[17] As a matter of public health, more than 100 countries have adopted unsparing language and dramatic, dogmatic warnings in their efforts to discourage cigarette smoking.[18]

Finally in 1966, more than 50 years later, states in the US began to enact some measures to curb smoking and other tobacco use through product labeling, public awareness campaigns, limits on advertising and restricted sales to minors. Doctors finally stopped appearing in cigarette ads and started counseling their patients about the undeniable health consequences of smoking. But this fatally slow process of reducing tobacco use gained traction only very recently – in the late 20th century – a lag that should give us pause if we're tempted to dismiss the slow public health responses to cholera or scurvy as a reflection of their pre-modern times. So far, in medicine's ability to respond to a widespread public health problem, little seems to have changed in the modern era.

In 2011, the US FDA announced new rules for cigarette warning labels that would mirror the European style that requires large, graphic images and warnings to appear on every pack sold in the United States and in every cigarette advertisement. These rules, drafted in compliance with the federal Family Smoking Prevention and Tobacco Control Act, were signed into law in 2009, but their future remains uncertain. The warnings were scheduled

to appear on US tobacco packs by September 2012, but a coalition of tobacco companies successfully sued the FDA, arguing that the new law would violate their free speech. Big Tobacco having won their injunction, the labeling requirements of the Tobacco Control Act remain on hold for now, and packaging in the US has not changed.[19] The text-only Surgeon General's warning first introduced in 1966 has not been updated in over 30 years!

So even today, having caught up after the delay between knowing that tobacco is a killer and proving it with science, our US commercial messaging does not make that reality clear to smokers. We need to push for change, a full paradigm shift, in the way we approach colossal public health challenges – not only within medicine but across multiple systems, most especially at the level of empowering individuals with accurate information and the authority to act in their own best interests. Big Tobacco is a most powerful and formidable adversary, able to resist the laws of the land with lobbying, to defy government regulations, and to shape the market forces that promote increased sales with bulk marketing and discounts. And now that public health reports and cultural changes have finally begun to reduce domestic sales, they are pushing to expand their sales in other countries around the globe – a strategy begun long ago, to globalize smoking.[20,21]

Borrowing Big Tobacco's playbook

Because diseases of tobacco use pose massive public health problems, our responses to it – both institutional and individual – offer an apt template for considering how we might deal with our most pressing current health crisis, diseases of food. Both the partial successes and the failures of the US response to Big Tobacco can help chart a path for pushing back against the dubious assertions and problematic practices of 'Big Food' and 'Big Soda'. If we pay attention to the lessons offered by smoking – across industries, institutions, and continents – we can do better this time, shortening our response time and saving both lives and quality of life.

Today Big Food and its cousin, Big Soda, have adopted strategies identical to Big Tobacco's. Indeed, when domestic sales of tobacco finally started to flag in the 1980s, tobacco companies' strategy for shoring up against the financial losses was to buy food companies.[22] The match was fundamentally well-suited: the tactics of disguising unhealthy products when marketing them for immense financial gain are Big Tobacco's well-worn specialty, so Big Food was a natural next step, and the current generation of food marketers is well-prepared. Like cigarette companies, food companies redirect responsibility for the diseases of food and soda onto those who use the products. Only when it became clear that it was a public relations problem with potential investors for food divisions to be owned by a recognizable tobacco brand did the companies later make efforts to separate the businesses and change their names.

As with tobacco, though, the problems with dangerous food are multilayered, systemic, and buried under misinformation – an array of obstacles to the average citizen who is seeking to make healthful choices and wanting access to safe food. Some of these barriers will remain insurmountable for consumers until we make systemic changes in our food supply.

Like all food questions, the story has its origins on the farm and continues through the supply chain, to mouths around the world. In the US, industrial monoculture grows a massive excess of corn, with subsidies from the US government, and many thousands of people are employed in its manufacture, promotion, and distribution. Soda and junk food products, filled with corn ingredients, are actively promoted to children and sold in schools, and they travel the globe with US troops. Promoting Big Food and Big Soda, salaried experts and advisory panels speak out to advocate on behalf of these non-nutritious calories, publishing reports that minimize the hazards of obesity, diabetes, hypertension, and heart disease while also serving as public relations vehicles, putting a favorable spin on the corn products in hopes of propagating favorable reports about ostensible health benefits.

Massive advertising campaigns promote junk food and soda products to young people especially, aligning products with Olympic health and fitness. As for cigarettes a few decades ago, the ads tout constant tweaks to food and soda products, always with the promise of less risk to human health. Meanwhile corporations wrestle to maintain perennial resistance to genuine disclosure, blocking revisions to clearer labeling that would lay bare the truth about their soda and food products. Bulk marketing discounts promote sales, and toys and promotional items target children with fast food and junk food products.

Because they have sprung from Big Tobacco corporations, Big Food companies know exactly how best to advertise toxic, addictive products. Tobacco primarily targeted teens and adults, but children were off-limits, protected by legal regulation; not so for Big Food and Big Soda, who target children deliberately, with their highly toxic, addictive, and disease-promoting products, using the same battle plan that was so effective for tobacco and the same tactics.[23] Openly reveling in the addictive quality of their products, food companies have boasted 'Bet you can't eat just one!' of potato chips, or named a sweet cereal for breakfast 'Crave'. Pushing junk food and soda, the game is the same as pushing tobacco but with even higher stakes, since everyone eats, at every age. Having learned their lessons well from tobacco, food companies too sponsor favorable research, lobby politicians, and use propaganda to spread doubt about the undeniably harmful effects of their sickening products.

To understand how to rally an effective response to the epidemic of diseases stemming from food, symmetry-seeking offers multiple paths and strategies, replicating those that have been proven to work well in other countries to constrain the popularity of cigarette smoking or to moderate the use of alcohol. Information is key, and accurate, unequivocal, and readily understandable labeling on packages is critically important. Without accurate information there can be no informed decision making. As on the cigarette pack, warnings about food must be displayed prominently on the front of the package, not hidden away on the side, and images and symbols speak louder than text. Advertising and promotion of harmful products should be restrained or discouraged, and products known to provoke disease should be subject to additional taxation. Public service announcements, campaigns to promote health, and local ordinances such as reducing serving sizes of soda beverages can each play a role.

On medicine and change

When disease looms large, we might expect our doctors and medical teams to be first on the field, well-suited to lead the charge and mount a coordinated response. So where are the doctors and health care providers, why are they not doing more about childhood and adult obesity? And where is the pushback to challenge the corporate grand deception? But as this chapter has demonstrated, the pace of medical advances is slow, and it is difficult to push for multisystem changes, even for known diseases with known causes, such as scurvy or cholera. The history of smoking shows us that some doctors with vested interests can even make themselves part of the problem, promoting unhealthful practices. Still more fundamentally, though, to look to our physicians to resolve our vast problem with diseases of food is to misunderstand what doctors do and misjudge the business of medicine.

Medical doctors are trained to recognize and treat symptoms and disease. They make their living by diagnosing diseases, counseling patients about their prognosis and helping to select the best course of treatments or management strategies for their symptoms and other complaints in the body such as minimizing their painful effects. These skills require knowledge and training, so medical education is lengthy and expensive, transferring protected secrets of the trade that are not to be lost, diluted, or even, historically, shared with those outside the field. To maximize their success in treating diseases, doctors follow simple, standard principles: that early detection is better than late, and that early treatments

are more effective than later efforts. Their desire to treat problems early therefore drives the popularity of screenings, increasingly common because they generate more opportunities for doctors to treat diseases while they are likely to be easier to treat. And the more doctors look for problems, ever broadening the dragnet of screenings, the more such favorable treatment opportunities they will find.

Unfortunately, while such an approach may be marketed as focused on wellness, it isn't: medicine, by definition and by its business model, inevitably focuses on disease, on diseases already present, with an ever-more laser-like focus from a specialist who will seek out any abnormality within that particular organ or body system, where disease might possibly exist. After years of studying the workings and disorders of the human body, doctors might understand the underlying course and causes of disease, but by the time a patient reaches the medical clinic with symptoms, it is all too late and time instead for the doctor to get to work, diagnosing and treating. Screening healthy subjects can lead to early detection and start the doctoring sooner, but if a disease is indeed present, detected, and worthy of treatment, wellness has already been lost.

Fundamentally, people hope not to become patients but instead to live without needing the help of doctors and hospitals. We do not dream of enduring plenty of successful treatments but the exact opposite: to be well and stay well, with as little medical intervention as possible, for as long as possible. We seek wellness, which is totally different in every respect from disease.

Upton Sinclair captured a truth we all know implicitly: 'It is difficult to get a man to understand something when his salary depends on his not understanding it'. And if we are to understand how best to make use of physicians' expertise, it is important to remember what their salary depends upon, which is illness and disease. Wellness – health – is not their bailiwick and never will be. The best doctors may know a great deal about the underlying causes of the conditions they treat, but they are not well-placed for effective prevention, and they would be out of business if it were their focus. And since the majority of fatal diseases today are the result of lifestyle choices – especially, of what we take into our bodies, whether from the farm or the factory – we are foolhardy to turn the task of maintaining our health and wellness over to doctors and the medical business, which can never succeed in keeping us truly healthy. Wellness is not their job, even if they would like it to be, even if hospitals advertise themselves as centers of wellness, and even if a doctor is supremely well-intentioned, ethical, skilled, and wanting the best for every patient seeking to regain health.

Some aspects of your personal health were decided before you were born. Equipped with the lens of symmetry-seeking, tuning in to your own body can help you and your care providers better understand what you see – how your body parts and body systems interrelate and what they are telling you about your best personal paths to wellness. If you are a health care professional who provides preventive care and guidance on wellness, these symmetry-seeking, whole-body questions are the place to start with your patients or clients. And even traditional, disease-treating doctors, including specialists, can support integrated care, and they will do best by their patients when they seek opportunities to practice whole-body medicine. The wide view is a good starting place, looking back to before the onset of symptoms, considering the whole, looking for patterns, and asking the symmetry-seeking questions about one's own body. Remembering how one part affects the others will give clues to the need for care in other parts of the body.

Even before the recent advances in genetics began to illuminate why, medicine had come to understand that inherited traits determine not only visible similarities across generations but inner similarities and risk factors for certain diseases. Symmetry-seeking sharpens our focus on the kinds of questions we should be asking to determine what some of those similarities might be, beyond the short list of family history questions standard on most medical forms. We can each learn a great deal from more thorough inquiry into the health and illnesses of our family members, and well-informed guidance from health professionals can hone our knowledge of what to look for. Looking upstream, you can broaden your perspective on the range of traits, vulnerabilities, and habits you may share with your family, including mental health tendencies, shared ways of family living, and traditions of food choice and cooking practice, which point and project us on similar trajectories towards shared health destinations. The more you know of your family's health profile, the better prepared you can be to steer your personal course away from predictable hazards.

Wellness is each person's own job – an inside job, for every individual. Just as your banker is not in charge of your personal finances, your doctor is not responsible for day-to-day decisions about health. For wellness, prevention is the key: enacting the behaviors that maintain health and avoiding known causes of disease. Medicine – the business of treating disease – will never be more interested in prevention than in diagnosis. To try to turn the medical establishment into the curators and purveyors of wellness is to place the fox in charge of the henhouse, just as pharmaceutical companies will never be the best protectors of drug safety and the food industry will never be the best defenders of accurate food labeling. Notwithstanding the good intentions of any one individual or corporation, their bread is not buttered on the side of our individual health. Sometimes our interests in health and theirs in profits dovetail, especially when a company's primary focus and marketing is towards healthful ingredients and production. But when our need for health diverges from food distributors' need for profit, that conflict of interest makes them one more fox in the henhouse.

To guide us towards achieving wellness, research – not only medical research but studies and data collection in all the many fields that touch human health – must focus on causes, not simply on honing doctors' ability to detect symptoms and signs after they appear. Armed with knowledge of an illness's origins, we can act long before any diagnosis or treatment, perhaps forestalling the need for either.

But if we are to prevent or avoid disease, we need to know how to stay away from what causes it, when we can. Now that we understand how foodborne infections occur – most commonly the result of eating raw foods of animal origin – we know to favor hot foods, and if eating meat, poultry or shellfish, to be sure that it is well cooked.[24]

It was research which taught us that smoking kills, generating knowledge about the dire effects of tobacco use, which public health campaigns could then disseminate, successfully persuading millions of Americans to stop smoking or avoid starting. Such knowledge has effectively reduced the percentage of the US population who smokes from more than 42% to less than 17% over fifty years.[25]

We have seen similar successes in environmental initiatives that affect human health, in city rivers cleaned up from their status as dumping grounds for toxic waste a few decades ago, or in the recent report that the hole in the ozone is actually shrinking.[26] In both instances, progress has occurred in response to science alerting us to a threat, organizations campaigning for public awareness, citizens expressing concern and support for change, and industry and government enacting policies to reduce pollutants: in the case of ozone protection, a 1996 ban on chlorofluorocarbons – also known as CFCs, gases used in aerosols, as cleaning solvents and in refrigeration units – in consumer products. The clean-up of US rivers was part of the impetus for the formation of the Environmental Protection Agency (EPA) in 1970.

Mapping patterns

To make policy changes, decision makers and policy changers need a weight of evidence, irrefutable data from multiple fields without confounding contradiction. In major public health decisions, much hangs in the balance. Fortunately, we do have topics with just such consensus among a broad, diverse array of experts, on many matters related to health, such as those which serve as case studies throughout this book. Much of our best knowledge about the behaviors, institutions, and substances which affect our

health comes from just such interdisciplinary teams of experts, in public health organizations such as the Centers for Disease Control in the US or the Public Health Service in the UK. Here professionals devote their careers to inter-systems questions of public health trends, looking for patterns and possibilities in preventive medicine, community or industrial health, advocacy, and policy making.

When we talk about 'your doctor', we usually picture your primary care physician, but some of the most important doctors whose work affects our lives are in public health. Typically, we never meet them, because these physicians may not work with patients, if they spend all their time doing field work, in research laboratories, or in the offices of organizations that tap their expertise as a resource for decision making. But they are well positioned to have a profound impact on our health – to put medical knowledge to work more effectively and efficiently than in the past, through the work of clinical health professionals and directly through individuals, by shortening the timetable turnaround from the days of scurvy and cholera and cigarettes, to the benefit of all of us.

Our public health and preventive medicine specialists serve critically important functions such as monitoring patterns of disease and administering annual flu vaccination initiatives, but resources are always limited in government, so political resolve will be essential if we are to mount effective, broad-based strategies for wellness.

Medical research, public health research, and epidemiology have already given us more than enough information and analysis to establish that highly processed calorie-rich junk foods kill. Now, making the best case for change in corporate practices, public policy, and, most important, personal choices for the prevention of obesity requires clear messaging from all sides that reveals the linkages and makes the connections between diet and disease. And symmetry-seeking can point the way to the clearest, most powerful messages – the most persuasive data – if experts from multiple disciplines, across widely varying fields, take the initiative to look outside their own silos and learn from one another, to apply symmetry-seeking principles to what they learn, finding out how the observed pattern in one system parallels patterns in others.

As both the London cholera and the Southern pellagra epidemics demonstrated, a prime tool for pattern-hunting in public health is mapping, because the geography of disease can paint a picture and point to possible causes. In today's obesity epidemic – fast becoming a pandemic – clustering of cases helps us to see its patterns through a different lens and begs for further exploration. Colorado, the mountain-high state, has the lowest obesity rate in the nation, while Louisiana, Alabama, and Mississippi, clustered at the mouth of the Mississippi River, have the highest.[27]

The geography of human obesity in the US follows the watershed of the river, from the cornfields of Iowa in the north to the Gulf of Mexico in the south, suggesting a strong link between land use and human health in this watershed. Researchers seeking the causes of obesity and its companion diseases will do best to look for systems with corollary geography: of nitrate runoff, poverty, dietary habits, soil, and water quality, characteristic industries and most common types of work in the areas where obesity is most common and most severe.

Slow systemic responses

We've seen how hard it can be to identify a disease and recognize its causes, and that even after medical science has found both an effective treatment and a way to prevent it, many decades may pass before medical practice, institutional policies, and personal behaviors begin to effect the changes necessary to manage, minimize, or eliminate a problem. A contained and straightforward condition such as sailors' scurvy or climbing boys' scrotal cancer can focus and hasten scientific success, but even in such cases, practical progress is still slow, with lives lost in the meantime. For a global challenge such as cigarette smoking or HIV infection, connecting the dots takes even longer, and policy and behavioral change

take longer still, even after all the science is in place. How much more challenging, then, is the quest to link diets and diseases whose causes spill from every pot and pan to dish and plate, intertwined with everyone's daily fare and touching all corners of the planet.

When seeking to apply the principles of symmetry-seeking and stepping back to look for broader perspective, searching for patterns and connections between food and health, we face infinite possibilities about the angle or level at which to set our lens. Nevertheless, given humans' inevitable, eternal, necessary interest in food, we now have millennia of human experience with the omnivore's dilemma of what to eat, centuries of science noting our body's reactions to certain foods, and now increasingly sophisticated data about those connections. Still more important, despite the overwhelming volume of that data and even amidst individuals' unique dietary needs and preferences, an impressively broad consensus emerges as to what kinds of foods can carry the vast majority of us further away from heart disease, cancer, diabetes, and dementia and towards longer, healthier lives. While research from year to year occasionally indicates that one food or another is better or worse than we had thought, the overall arc of nutritional advice has not changed in some time: to consume hearty amounts of vegetables, whole grains, fruits, and fish, and to keep intake of sugar to a minimum.[28]

An individual consumer may feel buffeted by conflicting nutritional advice, but many of the shifting edicts on particular foods derive from two main sources. Some come from vested interests seeking to promote a food as healthful with advertising, such as the 'Got Milk?' campaign – and in Chapter 9, we will consider the impact from media and advertising. These messages may actively seek to drown out widely accepted science to the contrary, even in some cases influencing public policy for USDA dietary recommendations or agricultural subsidies, towards food crops such as corn, which is not our best source of nutrition. Another source of news that a food is unexpectedly

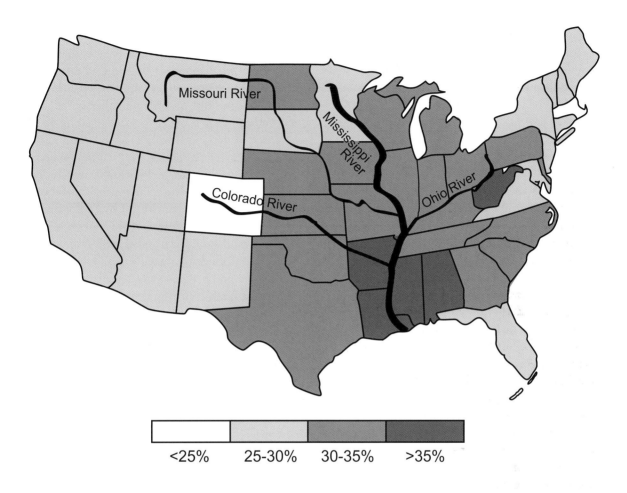

Figure 8.1

Prevalence of Self-Reported Obesity Among US Adults by State and Territory, 2016. Note how the rate of obesity is lowest in Colorado, in the head waters of the Colorado River that drains into the Mississippi, and increases steadily as you travel downstream. The corn-belt states are drained by the Missouri, Mississippi, and Ohio Rivers that together become the Mississippi, and with the exception of West Virginia which drains mainly to the Chesapeake Bay and Atlantic ocean, the highest rates of obesity are all clustered in the lower third of the river watershed – Arkansas, Louisiana, Mississippi, and Alabama. (Data from Behavioral Risk Factor Surveillance System, https://www.cdc.gov/obesity/data/prevalence-maps.html.)

detrimental to health comes from scientific discoveries about problems with how a product is grown or processed, usually due to industrial methods of production or preparation.

One by one, we are learning what some of the problems are in the foods we eat, many of which we address in this book. But there is much left to learn, especially when we consider the research indicating that an adult today, eating the same foods in the same amounts as his parents, is more likely to be overweight.[29] Clearly, we need some new perspectives beyond the simple advice to eat less, to understand what is causing our obesity epidemic, and symmetry-seeking offers a way of seeing with a fresh eye. It directs us to look beyond an isolated meat or fish product to consider what was the food that fed your food. Was it grass-fed meat or wild-caught (plankton-fed) fish? They will be rich in omega-3, a building block to be disassembled by your digestion and made available again for reassembly, providing a plentiful supply of key ingredients for healthy cells and tissues for your body and brain.[30]

Still, while we track the multifarious causes of current problems with our food supply and eating habits, it is basically good news to discover that there are no enormous surprises from the increasingly thorough research into food. We might wish for the novelty of some surprises – for difference, for the entertaining asymmetry of a grapefruit diet, a paleo or a vegan diet – but the boring reality of healthful eating holds steady to what we already know from our parents: to eat our vegetables and not too much ice cream. Ultimately, it is reassuring to realize that, notwithstanding some slippery slices of misinformation, we need not start from scratch on understanding what to eat, because we already know a lot. We will be busy enough figuring out how to conquer the challenge of changing our eating habits to act on what we know (also part of Chapter 9).

For the person who wants to maximize her odds of living a long life, with as much vitality as possible, still more good news lies within the grim statistics about these diseases of food. This chapter has revealed how slowly the great ship of medicine turns, even when danger is looming. Treatments and prevention for obesity and its companion health problems are apt to be a long time coming, requiring changes in the multiple industries discussed in this book, and in the meantime, we can expect these diseases to continue as the main causes of chronic illness and early death. But most of us have a great deal of say-so in what we eat, so we need not wait for the ship to turn. In the main, the science on what's best to eat is already clear. We don't have to wait for the Royal Navy to change its rations, and no need for cities to pass ordinances outlawing supersized soft drinks; we can

avoid them ourselves. Anyone paying a modicum of attention to health news understands the risks, so we know what to do, as individuals, to reduce our chances of disease.

Even while we work towards better mass-produced food choices, better public policy regarding food safety, and better access to healthy food and nutrition education – for everyone, across the socioeconomic spectrum – we can also start right away eating foods that lend themselves to better health now and better health later in our lives. On your next trip to the grocery store this very week, you can choose to skip that one box of snacks made with undecipherable, non-food, chemical ingredients and select something your grandmother would have recognized as food instead. In increments as we manage our own eating, we can invest in our current and long-term health and simultaneously be making a choice that is better for the planet's ecosystem, better for our city's waste systems, and much better for unborn children. Furthermore, every purchase of a product that is part of a more sustainable chain of symmetries is a vote towards turning the ship of interlocking institutions and policy – away from fatal food flaws and towards better alternatives for health and wellness. We can be on the lookout for such products to appear, such as Stonyfield yogurt's reducing its sugar content in response to customer health concerns.[31] In the UK, Public Health England has researched and published official guidelines to reduce sugar content in foods marketed to children, targeting the fare they eat most, such as breakfast cereals and fruit spreads.[32]

Knowing what is most advisable to eat is not the same as following that advice, however, as we all know. Healthy eating presents a formidable challenge to many of us, as evidenced by the multi-billion-dollar diet industry and by those statistics on heart disease. As we explored in the earlier chapters, the roots of our challenge lie in a problem of evolutionary mismatch, for we evolved to live in an ancient world of intermittent famine and drought, and our bodies were built to tolerate hunger and thirst. Some of the dimensions of this mismatch may be familiar, such as why we crave fats and sweets so endlessly, when large amounts of them are so bad for our health. But to understand why healthful habits can be a battle to establish, we must turn our symmetry-seeking lens back to the human body, to look at the form and function of a part we have only touched on thus far: the human brain.

References

1. World Health Organization (2017). Children: reducing mortality. Online: www.who.int/mediacentre/factsheets/fs178/en/.

2. Davis RJ (2012). Coffee is Good for You: From Vitamin C and Organic Foods to Low-Carb and Detox Diets, the Truth about Diet and Nutrition Claims. New York: TarcherPerigee.

3. Mineo L (2017). Good genes are nice, but joy is better. Online: https://news.harvard.edu/gazette/story/2017/04/over-nearly-80-years-harvard-study-has-been-showing-how-to-live-a-healthy-and-happy-life/.

4. Fryar CD, Carroll MD, Ogden CL (2014). Prevalence of overweight and obesity among children and adolescents: United States, 1963-1965 through 2011-2012. Atlanta: National Center for Health Statistics.

5. Ogden CL, Carroll MD, Lawman HG, et al. (2016). Trends in Obesity Prevalence Among Children and Adolescents in the United States, 1988-1994 through 2013-2014. *Journal of the American Medical Association*, 315(21): 2292–2299.

6. Ogden CL, Carroll MD, Kit BK, et al. (2014). Prevalence of childhood and adult obesity in the United States, 2011-2012. *Journal of the American Medical Association*, 311(8): 806–814.

7. Ogden CL, Carroll MD, Curtin LR, et al. (2006). Prevalence of overweight and obesity in the United States, 1999–2004. *Journal of the American Medical Association*, 295(13): 1549–1555.

8. Ogden CL, Flegal KM, Carroll MD, et al. (2002). Prevalence and trends in overweight among US children and adolescents, 1999-2000. *Journal of the American Medical Association*, 288 (14): 1728–1732.

9. United States Census Bureau. Small Area Income and Poverty Estimates (SAIPE) Program. Online: https://www.census.gov/programs-surveys/saipe.html.

10. American Medical Association House of Delegates (2013). Recognition of Obesity as a Disease. Online: http://www.npr.org/documents/2013/jun/ama-resolution-obesity.pdf.

11. Centers for Disease Control and Prevention. National Center for Health Statistics: Heart Disease. Online: https://www.cdc.gov/nchs/fastats/heart-disease.htm.

12. World Health Organization (2017). The top 10 causes of death. Online: http://www.who.int/mediacentre/factsheets/fs310/en/.

13. Rosenbloom M (2017). Vitamin toxicity. Online: https://emedicine.medscape.com/article/819426-overview.

14. Johnson SB (2006). The Ghost Map: The Story of London's Most Terrifying Epidemic – and How It Changed Science, Cities and the Modern World. New York: Riverhead.

15. Ali M, Nelson AR, Lopez AL, et al. (2015). Updated global burden of cholera in endemic countries. *PLoS Neglected Tropical Diseases*, 9(6): e0003832. http://doi.org/10.1371/journal.pntd.0003832.

16. Blondia A. Cigarettes And Their Impact In World War II. Online: http://www.calstatela.edu/sites/default/files/groups/Perspectives/Vol37/37_blondia.pdf.

17. Centers for Disease Control and Prevention (2012). Current Cigarette Smoking Among Adults – United States, 2011. Online: https://www.cdc.gov/mmwr/preview/mmwrhtml/mm6144a2.htm.

18. Action on Smoking and Health (2016). More than 100 countries now require graphic picture warnings on cigarette packs – UK goes further by requiring plain standardised packaging. Online: http://ash.org.uk/media-and-news/press-releases-media-and-news/more-than-100-countries-now-require-graphic-picture-warnings-on-cigarette-packs-uk-goes-further-by-requiring-plain-standardised-packaging/.

19. Bardi J (2012). Cigarette Pack Health Warning Labels in US Lag Behind World. Online: https://www.ucsf.edu/news/2012/11/13151/cigarette-pack-health-warning-labels-us-lag-behind-world.

20. Campaign for Tobacco-Free Kids (2017). The Global Cigarette Industry. Online: https://www.tobaccofreekids.org/assets/global/pdfs/en/Global_Cigarette_Industry_pdf.pdf.

21. World Health Organization (2013). WHO Report on the Global Tobacco Epidemic, 2013. Online: https://books.google.com/books?hl=en&lr=&id=hrIXDAAAQBAJ&oi=fnd&pg=PP1&dq=tobacco+markets+world&ots=xk72qjg5ik&sig=eky5Qy_JIZyG4lPr1oZXyeSBHQ#v=onepage&q=tobacco%20markets%20world&f=false.

22. Caplinger D (2016). A Short History of Big Tobacco's Fling With Food. Online: https://www.fool.com/investing/2016/09/23/a-short-history-of-big-tobaccos-fling-with-food.aspx.

23. The Fifth Estate (2013). The Secrets of Sugar. Online: http://www.cbc.ca/fifth/episodes/2013-2014/the-secrets-of-sugar.

24. Centers for Disease Control and Prevention (2018). Food Safety. Foodborne Illnesses and Germs. Online: https://www.cdc.gov/foodsafety/foodborne-germs.html.

25. Centers for Disease Control and Prevention (2016). Trends in Current Cigarette Smoking Among High School Students and Adults, United States, 1965–2014. Online: https://www.cdc.gov/tobacco/data_statistics/tables/trends/cig_smoking/index.htm.

26. Stierwalt S (2017). Why Is the Ozone Hole Shrinking? Online: https://www.scientificamerican.com/article/why-is-the-ozone-hole-shrinking/.

27. Centers for Disease Control and Prevention (2016). Adult Obesity Prevalence Maps. Online: https://www.cdc.gov/obesity/data/prevalence-maps.html.

28. Hamblin J (2017). New Nutrition Study Changes Nothing. Online: https://www.theatlantic.com/health/archive/2017/09/moderate-intake-of-things-linked-to-health/538428/.

29. Khazan O (2015). Why It Was Easier to Be Skinny in the 1980s. https://www.theatlantic.com/health/archive/2015/09/why-it-was-easier-to-be-skinny-in-the-1980s/407974/.

30. Hjalmarsdottir F (2017). 17 Science-Based Benefits of Omega-3 Fatty Acids. Online: https://www.healthline.com/nutrition/17-health-benefits-of-omega-3.

31. Kowitt B (2017). Stonyfield Gives Its Yogurt a Makeover. Online: http://fortune.com/2017/02/13/stonyfield-yogurt-sugar.

32. New Food Magazine (2017). Guidelines on reducing sugar in food published for industry. Online: https://www.newfoodmagazine.com/news/36537/guidelines-reducing-sugar-food-published-industry/.

Using our brains

A freak workplace accident in 1848 left Phineas Gage with a terrible head wound. An explosives engineer, he was injured when a long metal tamping iron accidentally ignited the explosive charge in a bore hole. The iron rod blew out of the hole and passed straight into Gage's left cheek, through the front part of his brain, and out the top of his skull. Amazingly, in spite of this dreadful injury, Gage lived. Not only did he retain a full range of muscle movements, but his memories were intact and his speech was not significantly affected. Curiously though, the injury led to changes in his personality: he became impulsive and explosive, and some said he was no longer himself. This historic case brought attention to the remarkable divisions of purpose in a human brain. Gage's accident had injured the front of his brain, which had an effect on his personality, but not the vital brainstem area that controls breathing, circulation and basic functions.

This chapter looks at the brain as a whole, in light of its embryonic and evolutionary heritage. The prefrontal cortex that Gage injured is the newer part of the brain, evolutionarily speaking, which provides us with the executive functioning that makes us different from other mammals. We also still retain the older parts of our brains that we share with other animals, aspects of our ancestral past that function as part of our survival repertoire. And when a new human brain is developing, a number of key factors affect how it will function in the years to come, ranging from genetic inheritance, nutrition, illness, and accidents to early life experience, as that brain learns to negotiate the world and its environment with the help of caregivers.

The brain as a layer cake

In earlier chapters on the body, we discovered that all mammals share the same body plan. On that common foundation, each species has also evolved over hundreds of millions of years with its own structures uniquely suited to specific purposes. And our brains, like our bodies, have evolved as well, adding new layers on top of much older layers laid down during an earlier evolutionary phase, so most aspects of our human brain exist in other animals as well.

In the late twentieth century, neuroscientist Paul MacLean offered a model for understanding the brain ('the triune brain') that grouped its structures into three main layers, each linking the stages of individual human development to our evolutionary development, in common with other animals.[1] While necessarily an oversimplification, especially now that we have more accurate and sophisticated knowledge about how the various structures of the brain work, this classic model points us in a useful direction.

A layer cake offers a pretty good analogy to the brain, built as it is with one section growing on top of another. Like the first layer of a cake, the brainstem is the foundation of the brain, rooted at the junction with the upper spinal cord (which serves as a kind of live cake stand, with its structures intimately connected to the brain). This base layer of

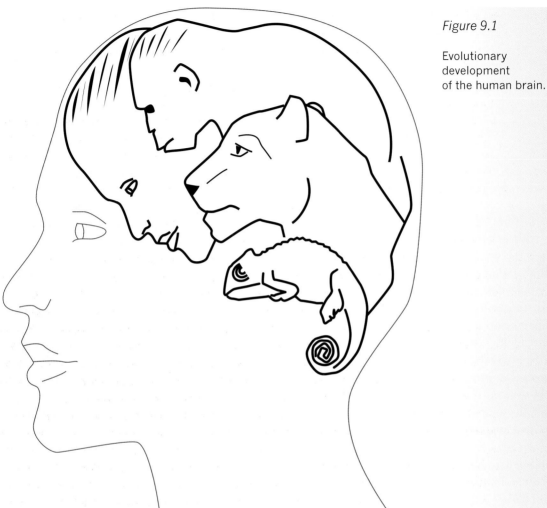

Figure 9.1

Evolutionary
development
of the human brain.

the brain, the brainstem, is the most primitive part, sometimes called the reptilian brain. It provides core functions for survival, and it is hardwired, which means that its patterns of functioning are fixed, varying little. The brainstem regulates basic vital functions that need to remain steady, like heart rate, breathing, and gut function, including maintaining the body's overall function in response to stress. It also processes incoming sensory inputs and the body's impulses toward movement, then informs the next level of the cake about emotions. The brainstem activates the survival responses of fight, flight, and freeze – not as thoughts, considered and acted upon, but as the body's direct reactions to danger.

Sitting above the reptilian brain is the next layer, the mammalian brain or limbic system, associated with motivation and social behaviors. Sometimes called the 'emotional' brain, with its functions very closely linked to the brainstem, this layer makes sense of the sensorimotor experience arising from the brainstem. Expressions such as 'my blood boiled' for intense anger or 'my blood ran cold' for extreme fear are based on the body sensations provoked by this layer of the brain, in response to sensory inputs. Many such expressions involve the heart, which can 'miss a beat' or 'sink' or 'break', depending on the person's inputs from experience. All these body-based phrases for feelings remind us that emotions manifest in the body, including sometimes in the foundational functions of our heartbeat or in our breathing. Emotion is motion, movement, in the body – not thought. A thought may spark a feeling response, but the event of the feeling is in the body. And positive emotions tend to lead us to move towards another person, where negative emotions tend to lead us to moving away, avoidance, or even hiding.

Within the mammalian brain layer are several structures, including the thalamus, the amygdala, and the hippocampus. The thalamus acts as a stimulus detective, whose job is to clarify whether what you saw in the grass was a stick or a snake. When the verdict is danger, the stimulus sets off our alarm switch in the amygdala, and triggers the brainstem to produce immediate survival responses. Immediately adjacent in this emotion-connected layer is the hippocampus, which is responsible for retrievable memory and past experience. The limbic system also processes sensory inputs and acts as a gateway to the higher layers of the brain – the cortex.

There above the mammalian brain sits a third layer, the cerebrum, with its two cerebral hemispheres. For both humans and other animals, here in the cortex lies their first real evolutionary opportunity for processing a more thoughtful, less reflexive repertoire of actions and behaviors, and the cortex is much larger in humans than in other mammals. Within the cerebrum lies the motor cortex, which drives coordinated muscle movements such as walking, running, and jumping as well as fine motor skills for preening and cleaning, writing and feeding, and the sensory cortex, which maps appreciation of touch and other sensations from the skin. Motor nerve pathways cross over from right to left and left to right, so the motor cortex of the right brain controls the muscle activity on the left side of the body, and vice versa. Special senses add

to the picture: the visual cortex, for instance, is the largest part of the brain, because for most mammals and humans, sight is so important.

Right brain, left brain

Like bilateral symmetry of the face and the body, the two-sided brain appeared in even the earliest evolutionary form of fishes and animals. Not only does having two hemispheres provide a kind of insurance – one side offering back-up capacity to the other side, in same the way that two kidneys are better than one – but the two sides also meet different and critical survival needs. Iain McGilchrist, a psychiatrist and neuroscientist, has written in depth about these complementary functions of the brain's two hemispheres.[2] To explain how the two sides work together to respond when faced with multiple, simultaneous demands, he suggests that we consider what happens when a small bird is pecking at a seed. To find the seed in the first place, the bird must pay close attention, to distinguish it from sand and pebbles; then it must focus on the task of aiming its beak at the seed, to break it open to eat. But in all these same moments the bird must also remain alert to other threats and opportunities as well, keeping an eye out for dangers such as a predator or an approaching vehicle, or even for a potential mate that might otherwise pass by unnoticed and be lost forever. Typically, the right brain scans the horizon, looks at the big picture, and includes the whole, while the left is narrowly focused on the task at hand.

In humans, too, we recognize a split between the two hemispheres, between more mental and more physical tasks, and knowing their dominant skills and developmental timeline is important for understanding a growing child. The right side, described as the 'experiencer', is intuitive and thinks holistically, in pictures – seeing the big picture, not just details. It receives experience as visceral, kinesthetic, and sensory, and it produces nonverbal communication through gesture and facial expression.

It lives in the present moment. The left side is more of an 'analyzer'. It thinks in language and looks for meaning-making, for cause and effect, interpreting the present moment in terms of the past. The left hemisphere can also then project forward, for planning the future. It sees the self as separate from others, making sense of its internal and external world as it plans and enacts daily living.

Not surprisingly, given the two sides' different interests, ways of being, and functioning within the two hemispheres, the right side dominates for the developing infant. Until around the age of three they absorb their knowing through doing and being, through facial expression, sing-song sounds of voice, and physical touch, rather than through the subtleties of language. In those early years, they develop their core beliefs about themselves and the world as a consequence of how they are cared for and how predictable and safe their environment may be, before their first memories start (around age 4). And by their seventh birthday, the two sides become more balanced, and verbal dominance becomes apparent.

Much wonderful writing is now available, such as McGilchrist's and Daniel Seigel's, which explores right-brain and left-brain functions, with many experiments supported by brain imaging studies, because we have reached an unprecedented and still-increasing level of clarity about how the brain is made, how it works, and why. Armed with this evolutionary asset of a two-sided brain and some knowledge of how it works, we humans have long recognized the broad categories of dominant strengths associated with each side of the two-sided brain, even before we knew about brain hemispheres. In recent decades the vocabulary of brain sidedness has even entered the vernacular, albeit with some over-simplification: the right-brain strength of the artist or poet in contrast with a heavier reliance on the left-brain for the accountant, mechanic, and doctor. All fishes, animals, and birds share the right–left body plan and right–left brains; even those tiny bird brains share our two distinct human imperatives, of focusing on a task while at the same moment maintaining a wider worldview: the left brain helps us to do, while the right allows us to experience.

Within this two-sided cortex, we finally come to the uppermost section of the brain, the most recent addition to the 'brain cake', belonging only to humans. If you hold your temples between your fingers, the prefrontal cortex is the brain tissue between your fingertips, and it is this topmost layer of the cake, which rests just behind our forehead, that makes humans so different from other mammals.

Prefrontal cortex

The prefrontal cortex helps us make thoughtful decisions. This section of the brain takes information – not only current sights, sounds, tastes, and smells but also memory of prior events and outcomes – and processes all these inputs, integrating the data. To recognize pattern such as we look for in symmetry-seeking, we use this part of our brain, the part that solves puzzles. The analysis considers all kinds of information, including postures, language, gestures, and actions; it compares a current challenge to past experience; it incorporates memory, knowledge, and emotion; and it tries to anticipate others' behaviors, all in an effort to set strategy and move towards effective resolutions.

Humans can pause and plan, taking time for thought, taking stock of a situation and using the prefrontal lobes to analyze and reflect, to draw on experience and consider multiple strategies or possibilities before acting. During this decision-making, both experience and emotion can come into play. An integration center, the prefrontal cortex is the executive boardroom, a place to take a moment for synthesis, to draw on all kinds of data from all available sources, to reflect and consider a range of possible actions or no action, from engagement to avoidance, according to the moment.

Humans are exceptionally equipped to communicate efficiently with one another because, in addition to speech and language, the prefrontal cortex brings a sophisticated system of social engagement possibilities that shape body language, posture, and facial expression. Contrast the expressive range of a human face with the typical reptile face, utterly mask-like and totally lacking facial expression; it is impossible to know what the alligator is thinking by the look on his face. A cat or dog is somewhat more expressive, but the human face brings a wealth of information, filled with both obvious expressions and a subtle range of variations, able to convey a great deal about an individual's prevailing condition, emotions, and state of mind.

When the brain is awash with adrenaline and cortisone, we don't think straight. Amidst a mammalian fight-or-flight response, we lose access to the analytical capacity of the prefrontal cortex. We thus become easier for others to manipulate when they turn up our fear, anxiety, and panic response. In his 'polyvagal theory' (based on his research on the vagus nerve, a tentacled system of nerve connections between the brainstem and the gut), Stephen Porges introduces the idea of social engagement systems and their importance in early childhood, personality traits, and mental health.[3] To be able to pause and plan, analyze and interpret events, and then execute a thoughtful, intentional response, we need to feel safe. We best learn how to achieve this state with self-soothing, even in the face of stressors, during our earliest life experiences.

Human development and secure attachment

A baby turtle, once hatched, will instinctively find its own way from deep in its nest to the sea, just by following the light of the moon, and a newborn gazelle is ready to follow its mother and the herd within minutes of birth. However, a human baby is critically dependent on caregivers to survive – not for a few hours, but for years. This long-drawn-out and complex process comes with significant costs, both in the elaborate care needed and the considerable risks to successful development. In the normal human fetus, the multilayered, bilateral brain has

formed by the time of birth but is still significantly undeveloped, such that, compared to other species, a newborn human more resembles a fetus arrived too soon in this world, incompletely formed. Incapable of independent movement and therefore utterly helpless, the human newborn's first imperative for life outside the womb is to bond to its caregiver and protector – usually the mother. With the tangible linkage of umbilical cord severed, other attachments replace it – ideally, in the moments immediately after birth – through the senses of touch, taste, smell, sight, and sound.

Born with an intact brainstem and its genetic inheritance, the infant will recognize its mother's voice and taste from its time in utero. Bonding is baby's vital first linkage, also called attachment – a very active process in the healthy infant. Attachment provides the baby with a sense of safe haven, born not only from feeding and tending to the infant's physical needs and safety, but also from closeness, care, and nurturing. Everything baby learns about the world is reflected from the caregiver, who ideally provides the loving base from which the infant and then child can explore, always returning for safety and security. Predictability in that relationship is key for emotional and social development, and any rupture in that relationship balance will require reparation and consolidation, and, like a wound, might heal with scarring and rigidity.

In both mother and baby, hormones support and enhance the bonds of attachment – hormones such as oxytocin, which regulates mother's milk production and baby's appetite. Porges describes a nerve pathway called the 'smart vagus', a neural circuit present from birth in mammals, which promotes feelings of safety in response to gaze, facial expressions, touch, and gesture. Between mother and baby, then, linkage grows in a complex, rhythmic dance of taking turns. A baby is attracted to its mother's voice, gesture, and facial expressions, and the mother equally anticipates the baby's expressions and sounds, responding with touch, baby talk, and gestures – all actively expressing and receiving emotions. The behavioral language of holding and being held, touching and being touched brings a silent dialogue of reciprocal calming and reassurance.

In managing a baby's mood and emotions, this give and take also serves to regulate the infant's internal conditions, including activation of the autonomic nervous system. The nerves that supply the muscles of facial expression, eye movements, voice control, sucking, and vocalizing also link with the heart, regulating and slowing the heart rate, as well as regulating the gut through the vagus nerve. So this social engagement system not only works to bond and socialize; it also influences the future ability of the infant to self-soothe, in later childhood. The calmer the caregiver and the more predictable at meeting a child's needs, the more the child will be calm and in balance, which provides optimum conditions for multiple aspects of growth and development. This relationship thus becomes a template for future relationships, and how we feel about ourselves relates to how we have been cared for and treated by others. If our limited world has been predictable and benevolent, then as our world broadens we will tend to view it favorably. A child born in a war zone, in famine, deprivation, or into domestic violence, or to parents who are emotionally unavailable – whether due to significant illness, addictions, or any situation that threatens attachment – is apt to have a very different, more difficult experience of the world. But a child who has had both overall good health and a secure attachment in this primary relationship is more likely to be full of energy and optimism, playful, and keen to explore, learn, and make positive relationships.

Beyond newborn and caregiver, attachment remains a co-creative dynamic that plays on throughout life in interpersonal interactions – pupil and teacher, athlete and coach, girlfriend and boyfriend – as the brain continues to mold its capacity. A mindful brain can scan the body and listen to its inner workings, becoming more aware of a racing pulse or rapid respiration. With this awareness and through attention to the breath, one can acquire better capacity to regulate and subdue the inner unrest or visceral commotion. Current research indicates that our chosen objects of attention, as well as the way we pay attention, activate neural connectivity and hormone messaging.

Balance and hyperarousal

As we have seen in chapters that look closely at the body, a healthy body is like Goldilocks' porridge, neither too hot nor too cold, balanced just right within a range of values. Inputs from experience activate ebbing and flowing in the body that barely register in the mind, rarely diverging too high or too low. Within this zone of balance, levels constantly adjust and correct to maintain a nearly constant optimal condition. This steady process of calibrating and recalibrating applies not only to hormones, as we discussed in Chapter 4, but to any number of metrics, ranging from body temperature, to hard to measure conditions such as mood. Our healthy bodies have a capacity to withstand assault and buffeting, only to keep returning to center, like the even-keeled boat that remains stable across a wide range of conditions. This balancing act also synchronizes with circadian rhythms, equipping the organism to optimize the cycles of light and dark, active by day and resting at night.

Disturbances disrupt the balance, however. When faced with a life threat or other danger, the inner systems switch the gears in the body engine's speed unconsciously, to prioritize survival above all else. In the well-known fight-or-flight responses, the alarm center in the amygdala triggers the hypothalamus, and the adrenal glands release adrenaline to stimulate mobilization. The heart races, breathing hurries. Digestion is not a priority in the face of a threat, so blood flow redirects away from the gut, and a person under intense stress may feel dry mouth, nausea, or an urge to move the bowels (and indeed we have a full range of colorful if crass expressions for fear based on those bodily experiences). When the body prepares for fight, the prevailing emotion will be anger, and the blood diverts to the upper limbs, bringing tension in the jaw, neck, shoulders, arms, and fists. In the flight response, the associated emotion tends to be fear, and blood is diverted more to the legs, in preparation for running away. Both fight and flight are high-energy states, or 'hyperarousal', which may bring feelings of panic, alarm, and anxiety. At higher levels of hyperarousal, the frontal cortex switches off, because this is not a time to be thinking your way out of trouble. The body wants action, and it prepares for the actions it knows best.

A third state of hyperarousal paradoxically resembles non-arousal: the freeze response, like a deer in the headlights. This reaction can appear during overwhelming terror and powerlessness, such as when young children who cannot fight or flee face great danger or high stress. This frozen state is primarily associated with reptiles, who are limited in their repertoire of stress responses; if you handle a pet snake and it feels threatened, it might react with sudden stillness and stiffness, shutting down. The similarity between a stressed baby and a reptile is not a coincidence but rather a throwback to our developmental heritage. Like reptiles, babies have only their most primitive brain layer adequately developed and ready to call on; they cannot even regulate their body temperature yet, the way other mammals can. Under threat, then, the reptilian layer of the brain takes charge, slowing the baby's heart and respiration, a response which can be dangerous. Even as adults, we might suddenly freeze in place when faced with extreme stress, our coping systems failing, quite unable to muster any other kind of response. Unfortunately, while temporarily decelerating the heart and slowing the breath may work well for reptiles, in young children it can compromise basic vital functions and even prove fatal.

As the brain grows in the first three years of life and motor centers develop, neural vineyards of the brain sprout with new growth, and corresponding novel skills appear: sitting up and crawling, early language, standing and walking, and finally running, leaping, and jumping. As the infant, then toddler, sets out to explore, the optimal conditions are secure attachment and safe surroundings. Psychiatrist Daniel Siegel's research into the human brain and interpersonal relationships asserts that the single most important requirement for optimal mental health and neural development is healthy bonding and secure attachment during early life.[4-6]

Like the circuit boards of the earliest computers, the newborn brain has only a few basic functions and a very limited operating system. Unlike computer circuits, though, brains are alive, with nerves ready to grow and divide, branch and reach out, spreading like creepers to make turns and connections in all directions. Within attached relationships, interactions drive neural activation and model pathways that, once sketched out, are reinforced and established in patterns that will form a person's neural templates for life. 'Nerves that fire together, wire together',[7] because tendrils that touch can reinforce one another to promote and propagate new growth. Nerves without attachment wither and die.

Just as it is easier to see our human symmetries in external body parts but not so easy in our internal organs, so it is easier to see progress in a toddler's motor skills like standing or walking but harder to note the less obvious skills of emotional development and personality. Physical skills and emotional resources build simultaneously, however, in the dyadic dance between caregiver and infant. And as a child's growing neural circuitry blossoms, new delights of responses echoed by caregivers lead to more engagement, interaction, and social interest, and finally towards a full and versatile emotional balance of personal and interpersonal resilience. Receiving soothing is, for example, the first step towards learning how to self-soothe. And the keys to wellness and good health turn out to be among these skills that can appear for the first time during infancy: a capacity for self-awareness and some degree of control over the internal workings of one's own body – even over some of one's organs' functions, since they are so closely linked to our emotional experience.

Mental and physical health

As with the healthy body, so with the healthy mind, which is resilient and stable within its Goldilocks range, rather than labile. A sturdy mind is able to self-correct, to right the ship and go with the flow, and a baby mimics its primary caregivers, tending to reflect and assume parental behaviors and acquire their skillsets. Not only does optimal attachment therefore foster resilience and confidence, but research suggests that it also plays an active role in shaping the fabric and circuitry of the growing brain. Secure attachment is the foundation for mental health, for emotional stability and a balanced personality – which together with inborn physical traits is also the foundation for physical health. Seigel's work on secure attachment confirms the importance of the initial conditions that set a trajectory for learning and psychological development, and independent evidence indicates that not only mental health but also vulnerability to physical illness is related to events and circumstances in the early years of life. Another perspective comes from the study of adverse childhood experiences (ACEs), recorded in a large CDC/Kaiser Permanente study that followed 17,000 patients. Robert Anda and Vincent Felitti were able to demonstrate an overwhelming correlation between ACEs and poor health outcomes, including increased risk not only of alcoholism and depression but also of cancer and heart disease.[8]

Early childhood stressors from adverse family and social factors thus have consequences even beyond the lack of a safe, secure, and sheltered platform for children and their interfering with attachment. Chronically recurring adverse events deny a child consistent loving care, and without sufficient experience of being cared for and soothed, a child is more likely to lack the personal capacity for self-soothing. Early deficit mirrors and predicts later deficit. With this skillset blocked, the zone of possible social engagement narrows, so that children are more easily dysregulated and may have difficulty settling in school, may be bullied, or just don't fit in. Learning is then problematic when your focus is elsewhere, and forming friendships is hard when you don't feel safe and maybe don't trust people.

In addition to interacting with other people, children must also be able to relate to themselves. Ideally, they learn to interpret internal sensations such as aches and pains, indigestion, tightness in the chest, or discomfort in the pelvis. A resilient child has the capacity and confidence to analyze such internal signals without being fearful, to consider triggers that might have invited the problem, reflect on prior experience, and remember what seemed to help the last time. Lacking secure attachment, the growing human brain fails to acquire full versatile form and function, failing to develop the capacity for analysis and decision-making, so the survival-based mammalian brain takes the wheel.

Mindfulness and meditation

Having studied the importance of positive caregiver–infant interaction, Seigel recognized strong parallels between patterns of positive outcomes after secure attachment and the benefits reported from mindfulness practices. According to Jon Kabat-Zinn, scientist and creator of a practice called 'mindfulness-based stress reduction', mindfulness is 'the awareness that emerges through paying attention on purpose, in the present moment, and nonjudgmentally to the unfolding of experience moment by moment'.[9] Meditation-based mindfulness has become popular for stress reduction, for cultivating wellbeing, and for expanding personal resilience.[10] The parallels Seigel spotted showed symmetry at work, as he compared the breathing patterns of mindfulness meditation to those of a mother and child: just as oscillations of the breath linked a caregiver with an infant in arms while soothing the child, deliberate regulation of the breath helped to integrate and align mind and body.[4,5]

For familiar activities such as sports, practice is essential to improving one's game. Whether throwing a baseball, swinging a golf club, or landing a jump, practice develops the skill. Focus, concentration, commitment, and repetition reap rewards in shaping movement patterns and muscle memory to establish a consistent, reliably effective action in sport. Better to start young when the body systems are flexible and adaptive, knowing that age and injuries will limit your range and stiffen your reach. But it is never too late for new beginnings. Targeted, persistent repetition with the right kind of coaching works the same way with a musical instrument. Eventually, after enough time spent working the piano keys, the brain is no longer simply directing the body consciously, because practice patterns develop neural pathways – brain–body integration – and the deliberate repetition is replaced by seemingly effortless flow that enhances this or any personal skill. For skills and abilities not acquired young, deliberate attention with an aim of mastery turns out to be a potent learning strategy at any time of life.

As a musician periodically needs time to tune – a little quiet spell to listen to the tone and adjust a musical instrument for resonant harmonious sound, for better playing – so every individual needs time to still the commotion of a busy day, attend to the body, and tune the mindful brain in a practice of awareness. The intentional practice of paying attention in the present moment and a decision to master this practice as an adult can bring remarkable rewards, particularly with the help of bodywork, and meditation can help us to acquire skills of enhanced awareness and self-soothing.

Mindful attention to the body, when nurtured, quiets ancestral brain layers and empowers the prefrontal cortex, enhancing health in many ways, both directly and indirectly. In addition to evoking calm and a sense of greater wellbeing in the present moment, mindfulness cultivates improved concentration and focus, allowing for more effective decision-making in any area – including personal health, which demands deliberate decisions every day. Mindful awareness practice can promote health by improving the linkage between mind and body, helping to harness and integrate organs and appetites. When we gain greater control, such as the ability to slow the pace of rapid breathing, deepen shallow breaths, and thereby slow and quiet the racing heart or pounding head, we can bring attunement and an increased sense of wellbeing, even strengthening the body's immune system.[11]

Health is balanced in a middle ground, and ill health lives beyond normal limits, in imbalance, whether in our bodies, our minds, or our interpersonal relations. If that linkage between parts and systems is impaired – when the connection to facilitate flow and balance are missing – we move toward detachment or rigidity. This realization that linkage and connection lead towards better health prompts us to tune the mind into better balance too, building connections within the body and between the mind and body, allowing us to cultivate and foster better connections wherever they are needed.

Early learning and what we believe

Just as early influences on the body are the most powerful impacts on one's lifetime health trajectory, so early influences on the mind set a course for how one learns and reacts for years to come. If the landscape of our mind is like a giant whiteboard, with each new lesson written in marker ink, then early learning is bold, written first on the clean board, large and strong in the open space. The message becomes more indelible if we hear it repeated, rewritten again and again on the board of our minds over the years, and it then becomes harder to erase or write over. New information has to find a corner and fill in spaces between what is already written.

And much of our early learning has to do with food and eating. The parents who feed us build our basic assumptions about eating, our family habits and assumptions about food and nutrition, imprinting onto our minds not just written with ink but engraved in granite – hard habits to break. Feeding punctuates every day with comforts and rewards, and some foods have special status from the beginning: milk, for instance, is baby's first food, and most of us learn to drink our milk, and that milk is good for you.

While it was true when you were a baby that milk was good for you, times change, and the body's needs change. The adult human does not need milk in the diet any more than does the adult sheep or cow; it was the lamb and the calf that needed suckling on calorically high-octane milk sugar and milk fat. But advertising takes full advantage of our primitive brain to exploit those first-learned, imprinted truths, even when some of them are no longer valid. Because agribusiness is about business, not about health, the successful 'Got Milk?' campaign reads as 'Got Money?' to the producers of milk, so they market it to adolescents and adults as if it were healthy.

This power of advertising was another surplus from the war years. Communications technologies grew exponentially during World War Two, to be used with military discipline: passing down orders; filtering, censoring, and intercepting messaging. Beyond the military, pro-war propaganda maintained civilian order and morale and promoted necessary efficiencies. After the war, those capabilities and communications habits developed for wartime were then retooled and transformed to create the US advertising industry (whose rise in the 1960s was depicted in the TV series *Mad Men*, in the 2000s). Taking full advantage of the wide range of new media outlets as they arose, the mass-produced telecommunications of radio, television, and film launched a national scale of outreach, influence, and press for business products. Promotional sound bites and advertising jingles thus entered the daily language, and over the last 70 years, these advertising slogans have risen to become ubiquitous and intrusive – unwelcome but constant companions, appealing more to our reptilian and mammalian brains than our prefrontal cortex.

Knowing what to believe

How do we know what to believe, when we are told that something is good or bad for us? For myself, my prefrontal cortex and I evaluate the evidence and its sources, considering whether the information fits with what I think I already know about the world. Some messengers are more or less reliable than others. Your mother, father, or partner, another family member, or a friend might be trustworthy and have your best interests in mind, and we get to know which of them is more and less well-informed about particular topics.

A health professional such as a nurse, pharmacist, therapist, or doctor should have more valid and up-to-date information than others about health matters, as long as there is no financial reward at stake in passing on the message – though such conflict of interest is becoming increasingly difficult to ascertain. Information from newspapers, magazines, radio, TV, and the internet deserves scrutiny, because what passes an editor's filter as newsworthy or catches a producer's eye as likely to bolster viewership may or may not also meet the test of reliable information, even if skilled presentation renders it more believable. Paid announcements deserve more than a pinch of skepticism – advertising may contain honest information, but since its express purpose is making money for the advertiser, ads carry not even a pretense of objectivity, so we do well to proceed with caution and to look for affirmation or contradiction of their claims from reliable sources. To be acceptable, new information should have a verifiable source and facts to substantiate its claim, as without such evidence it is no better than common chatter and speculation.

Fresh and cunning means of promotion for market share have made it harder to ferret out the facts from the mountains of health data, tactics that take full advantage of the confusion produced by the data deluge. Massive marketing budgets can also now use another dimension of contemporary data stockpiles – personal data profiles – to target advertising messages specifically to vulnerable consumers with health care needs, without regard for truth or accuracy. Promotional pieces that are difficult to distinguish from factual or objective news now appear alongside news articles, even in reputable publications. And advertisers know how to appeal to our ancestral brains and bypass our best decision-making with products that answer the very fears and stresses their advertising provokes.

Repetition reinforces messaging, writing the jingles again and again on our brain boards, and catchy phrases stick in the mind. Images and associations with a popular personality or activity provide linkage, and multiple hooks hold messages in our heads – whether welcome or unwelcome. And promoters borrow credibility from various organizations with professional-sounding names, claiming their vigorous support and implying a broad consensus among experts, when in reality the enthusiasm reflects only the size of the financial incentive for endorsing the featured product.

Symmetry as a test of truth

For us to accept a new piece of information, it generally must resonate in some way with our prior knowledge, not conflicting too strongly with what we hold to be true and thoroughly tested. And to pique a new ear, the message needs to be relevant to the listener's life, ideally carrying some potential for real, important, and personal benefit.

When deciding whether to accept new information as a fact and true, we consider it in light of what we already know, or think we know, to see whether it fits. Implicitly, this comparison already applies the principles of symmetry-seeking, as the notion of 'fit' conjures the image of a shape that does or doesn't match its intended slot. Has the puzzle piece found its place in the whole? Does a new idea match, balance, or compensate for one similar to it, in our existing structure of understanding? If, even after backing up to consider the bigger picture of what we know, we cannot find a place for this new knowledge within the pattern of anything we have already resolved to be true, we will reject it as false. The new information most likely to pass the test, then to qualify as truth, will fit somehow into our existing framework, like the puzzle piece we didn't know was missing until we found it. A new idea that passes the test and gives us an 'Aha!' sense of discovery will give clearer meaning and structure – symmetry – to balance, support, or counterbalance facts we know to be true.

If we are to accept and make use of it, new information is not just thrown in some corner of the brain or piled in a cognitive heap, burying, concealing, and obscuring what we already know. Rather, we fit it in, using pattern and symmetry actively as tools for organizing the human mind and memory, building connections and making meaning.

This practice of symmetry-seeking, then, you need not learn from scratch, because you do it all the time. The principles we offer and apply in these chapters are guides to some specific ways of using and honing the ways you apply the method – towards wellness and awareness of sustainable big-picture systems. Symmetry-seeking methodology provides a structure for adding new elements of knowledge to form an integrated framework that is easier to acquire, to retain, and to recall, when you meet new information. Symmetry-seeking can also help resolve episodes of cognitive dissonance. It prompts us to take a fresh look at something we already know about, but to move it around into a new context and to revisit it, finding another way to fit it together to make more sense or provide a better picture, one that resonates better than before.

Habit and games

Once we have applied our best symmetry-seeking tests to filter and accept or reject incoming knowledge, once we have drawn meaning and made decisions, it is time to act. In a conversation about health, this means it is time to eat – as we do, several times a day. Among the handful of proven practices known to benefit our short- and long-term health – including learning mindfulness, as explored in this chapter, or staying physically active – the science is absolutely clear that our food choices are the lynchpin of wellness.

Unfortunately, however, knowing what's best to do fails to translate directly into doing it. With eating, especially, we all know how difficult it can be to follow through on even the simplest knowledge of what foods are best for our health and which to avoid. It may or may not comfort you to remember that you are not alone in the challenge to eat more healthfully: researchers on willpower have found that wanting to eat more nutritious foods and less junk food is one of the most common – and most difficult – trials of self-discipline. It turns out that some of the conditions of dieting directed to weight loss, for instance, are a perfect set-up for failure, since one of the boosts to self-control – that same self-control required if we're to say no to a sugar-laden snack – is, ironically, an energizing boost to blood sugar. A team led by social psychologist Roy Baumeister discovered the connection by accident, in an experiment during which test participants received milkshakes during their breaks between challenging tasks. The research design included the shakes for their indulgent deliciousness, but it turned out that their effect on task performance lay in their calories (a finding that the researchers were able to test and put to further use in future experiments).[12]

Despite the difficulty of the challenge of eating well, health psychologist Kelly McGonigal offers good news in *The Willpower Instinct*, in a chapter she calls 'Your Body was Born to Resist Cheesecake'.[13] Comparing cheesecake to a saber-toothed tiger, McGonigal notes how the two are similar but not the same: a large mammal about to pounce posed a more frequent threat to life and limb back when our species' brain was evolving, and the danger triggered, appropriately, a fight-or-flight response. But the current largest threat to life and health – including limbs, such as arthritic knees or nerve damage from diabetes – is now the cheesecake. And our brains have prepared us for this moment too, if we learn to enact what McGonigal calls the 'pause-and-plan' response, enlisting the prefrontal cortex, to use the frame described in this chapter. And scientists are hard at work finding ways to make that challenge more manageable, to help us increase our reserves of willpower and self-control so that doing what we want to do – following through on what we have decided is best – can be easier and more achievable. McGonigal offers some of those strategies, and journalist Charles Duhigg presents

a thorough investigation of the recent scientific research in *The Power of Habit*.[14] In a happy synchronicity for our health, two of the practices that support better self-control for eating are strategies that also benefit health directly: mindfulness and physical exercise, both of which lower stress, calming the fight-or-flight response to allow us to pause-and-plan.

One dimension of the research on habit-formation, with numerous emerging applications to health habits, is what's broadly known as gamification. Recognizing that the goal of a long, healthy life involves countless instances of deferred gratification, some people are turning the process of eating well and exercising regularly into games, rewarding pleasure centers in the prefrontal cortex with intermediate wins for daily and hourly achievements. Tapping electronic technologies to aid in the effort, developers are creating apps and gadgets that give rewards for success, ranging from simple points in the game to prizes modeled on traditional video games with fantasy characters, leveling up, and unlocking achievements. Many games leverage the social dimension of motivation to spur participants' health efforts by allowing players to compete with friends or strangers, such as comparing the results of their automatic fitness monitors with others' workouts.

Not coincidentally, there is controversy around the idea of gamification precisely because of its associations with commercialism – with potential misinformation emanating from vested interests.[15] Jane McGonigal, for example, a game developer and researcher closely associated with the idea of 'gamification', has developed marketing games for major consumer corporations including McDonalds, Disney, and Nike, but is perhaps best known for creating a wellness app called SuperBetter, based on a game she created to salvage her own health when she was failing to recover from a concussion. In a curious coincidence of 'similar but not the same', Jane is the twin sister of Kelly McGonigal, the health psychologist, and they share a symmetrical interest in finding practical means to help individuals increase health and wellbeing. Jane disavows the term 'gamification' precisely because of its associations with manipulative commercialism, but – in a world where we spend more than 3 billion hours a week playing electronic games – she sees gaming and games, broadly conceived, as powerful tools towards solving major problems, whether for a person or in the world more generally.[16,17] The critical ingredient for success in producing real-world change is engagement: the person using game techniques needs to be intrinsically motivated by the end goals of the game. Notably, this same ingredient dilutes the potential ethical problem of commercialism, as the approach she proposes depends on personal agency: that we engage in games that move us consciously towards our own real and worthy goals, not those chosen for us by others who stand to profit from our choices.

In this broader conception of gamification, a playful or game-like approach is arguably the foundation of most strategies for achieving long-term goals, by breaking them down into manageable, day-to-day tasks. The humble Pomodoro method, of earning checks on a chart after successfully completing blocks of time on a task, is a low-tech game, of a sort, with oneself. It is also a tactic with proven success, and more than a few mobile apps, including popular meditation apps, have successfully built on the reality that we never outgrow some level of satisfaction at earning a gold star for good behavior – better yet, a row of them.

Ultimately, if they are to be effective, the crux of any of these approaches to better eating, better health, or sustainable practices more generally is this notion of agency: of deciding for ourselves what to do, rather than having it decided for us by vested interests who have neither our wellbeing nor the planet's as their priority. In the language of games, we could say that we must learn to play or else be played. The symmetry-seeking approach is a satisfying way to look at new and old problems, to process information, but its ultimate usefulness is in its application to daily life and to predictable challenges, both small and vast. The strategies of symmetry-seeking will help you evaluate data and make better guesses in the face of incomplete datasets. The challenge, living in an evolutionary body that has sacrificed optimal tail-end function for supreme cranial performance, is to make the most of these glorious brains of ours not only in determining the truth and choosing a better course of action, but in finding the discipline to face down the tiger that is only and always, every day, a cheesecake.

Making the connections

When we view the brain and its functions through the lens of symmetry-seeking, the most visible common thread is all about connection. The layers and hemispheres of the brain connect not only with each other but with all the systems of the body, and for our neural circuits to wire together for proper life-long function, we need connection to other human beings – interpersonal interaction, ideally starting with secure attachments in the family, between baby and parent. Mindfulness practice offers connection with oneself, promoting the self-awareness that allows for stress reduction and self-soothing. An ability to connect new information with what we already know allows us to re-evaluate both, amidst the mayhem of incoming data, and symmetry-seeking tools help us decide what information to trust. The playful spirit of a game helps many people then connect their long-term goals and hopes to daily behaviors that will contribute most to their health or happiness. Symmetry-seeking itself is ultimately about making connections, as we look for sameness, similarities, and linking one with another.

In this connectedness of touch, tuning, and balance, biotensegrity is the tangible push-pull architecture of physical linkages between tissues and throughout the body. In the world beyond our bodies, social cohesion functions as a kind of linking architecture of relationships, where our brains find the knowledge and connections to others that we need to function as effective human beings. Our next chapter will consider some of the broadest applications of the symmetry-seeking approach: to understanding the cultural systems, institutions and industries that define and determine most about our lives and our health, as we decide where we fit within those complex, counterbalancing structures.

References

1. MacLean PD (1990). The Triune Brain in Evolution: Role in Paleocerebral Functions. New York: Springer.
2. McGilchrist I (2010). The Master and His Emissary. New Haven: Yale University Press.
3. Porges SW (2014). Clinical Insights from The Polyvagal Theory: The Transformative Power of Feeling Safe. New York: WW Norton & Co.
4. Seigel DJ (1999). The Developing Mind: How Relationships and the Brain Interact to Shape Who We Are. Guilford Press.
5. Seigel DJ (2003). Parenting from Inside Out. Jeremy P Tarcher.
6. Seigel DJ (2007). The Mindful Brain. New York: WW Norton & Co.
7. Hebb DO (1949). The Organization of Behavior. New York: John Wiley & Sons.
8. Felitti VJ, Anda RF, Nordenberg D, et al. (1998). Relationship of childhood abuse and household dysfunction to many of the leading causes of death in adults. The Adverse Childhood Experiences (ACE) Study. *American Journal of Preventive Medicine*, 14(4): 245–258.
9. Kabat-Zinn J (2003). Mindfulness-based interventions in context: past, present and future. *Clinical Psychology: Science and Practice*, 10(2): 144–156.
10. Bhasin MK, Dusek JA, Chang B-H, et al. (2013). Relaxation response induces temporal transcriptome changes in energy metabolism, insulin secretion and inflammatory pathways. *PLOS One*, https://doi.org/10.1371/journal.pone.0062817.
11. Davidson RJ, Kabat-Zinn J, Schumacher J (2003). Alterations in brain and immune function produced by mindfulness meditation. Psychosomatic Medicine, 65(4): 564–570.
12. Tierney J (2011). Do You Suffer from Decision Fatigue? Online: http://www.nytimes.com/2011/08/21/magazine/do-you-suffer-from-decision-fatigue.html
13. McGonigal K (2013). The Willpower Instinct: How Self-control Works, Why It Matters, and What You Can Do to Get More of It. New York: Avery.
14. Duhigg C (2013). The Power of Habit. London: Random House.
15. Fuchs M, Fizek S, Ruffino P, et al. (2014). Rethinking Gamification. Meson Press.
16. Feiler B (2012). She's Playing Games With Your Lives. Online: http://www.nytimes.com/2012/04/29/fashion/jane-mcgonigal-designer-of-superbetter-moves-games-deeper-into-daily-life.html?_r=0.
17. McGonigal J (2012). The game that can give you 10 extra years of life. Online: https://www.ted.com/talks/jane_mcgonigal_the_game_that_can_give_you_10_extra_years_of_life.

Confluence over conflict

On January 17, 1961, Dwight D. Eisenhower delivered his final presidential address to the American people, famously warning against the dangers of war-minded vested interests driving US policy and business. Before World War Two, he explained, there had never been a standing arms industry in America, but the imperatives of the war had created a vast defense industry, which was becoming permanent amidst the Cold War. Recognizing the scientific advances achieved in the service of the war effort, he also foresaw the difficulties of balancing the requirements of peace with the prerogatives of a military that dominated an enormous sector of the economy. Eisenhower issued a caution about the role of scientific research and discovery, that science not be dominated by the 'unwarranted influence' of the 'military-industrial complex', and, on the other hand, that the academy not capture and limit public policy either. His recurring theme was the necessity of balance, and he invoked the perspective of time:

> As we peer into society's future, we – you and I, and our government – must avoid the impulse to live only for today, plundering, for our own ease and convenience, the precious resources of tomorrow. We cannot mortgage the material assets of our grandchildren without risking the loss also of their political and spiritual heritage.

Today, more than fifty years later, we hear this warning as a call to sustainability, and Eisenhower's caution resonates more loudly than ever before, as many of the dangers he foresaw have come to pass.

The challenge of complexity

The systems that support and sustain our way of life, which we in turn support and sustain, are interdependent like architectural elements in biotensegrity, such as we saw in Chapter 1. Our institutions are many, woven into complex, interlocking infrastructures, with hierarchies of parts within parts, and linkages that scale from personal habits to the whole planet earth. Although many of us try to stay alert, it is impossible to know about all the factors that affect our daily lives. The colossal magnitude of the task is far beyond any one of us, and, by their very nature, complex systems are difficult to understand. Certainly, the structure of the US military, the many branches of our multitiered governments, or the forces that drive our economy – from supply and demand to tax incentives, resource limits, and marketing – or even one industry within that economy, are beyond the full comprehension of any one person.

Even our familiar human body is not entirely knowable. As we have explored in this book, however, when we step back for a fresh look, the body shows itself to be rich in orienting symmetries, which give us enough understanding to make sense of it and to guide us in how we tend to it. We need not comprehend every detail to understand the whole. That beneath our skin we all share this same body plan, made over and over again, offers still broader perspective from other human bodies and, for most of the plan, other species. Our life form has survived and thrived from primordial beginnings. We found a robust method of procreation, growth, and development by employing the well-tested, well-honed instructions of Hox master genes, in a toolbox not reserved exclusively for human use. And of all the signaling systems in that ancient cellular toolkit, hormones are the most potent and powerful, and just the same for all. Hormones are not only masters of growing and growth but drivers of our every cell and organ, fueling appetites and shaping behaviors.

Making sense of complex systems takes time. It took nearly 50 years after DNA was isolated for Watson, Crick, and Franklin to define its double helix structure. And its molecular workings made sense only when its structure was revealed and its pattern and symmetry defined. Like DNA, the insides of the human body are intrinsically complicated, but pattern and symmetry have helped to make it more understandable.

Like our bodies, with their distinct structures and systems, science too is divided into many separate parts with different experts, each a specialist in one part or another, producing new knowledge in relative isolation, publishing in professional journals to only their own fellow specialists. Separate fields of research, with firm boundaries between them, serve to contain specialty knowledge within silos, and although collective experience is constantly expanding communal knowledge, only a small part of that vast learning can be known by any one person. This established separation illustrates the misfit between the organization of scientific knowledge and the challenge of addressing the large-scale, multifaceted problems of today's world, where each part affects the whole while the effectiveness of the whole, in its totality, also affects the parts.

The explosion of scientific knowledge in the mid-20th century brought both the beginnings of hyper-specialization and the first glimmers of deep cooperation across disciplines, and we are currently in the historical moment that challenges us to find an apt model for the big, big picture – of the interlocking systems that shape human health, of how they have evolved to function and how they can work much better, more sustainably, to support human life and the planet.

The mindset of war

Efforts to make sense of these complex, interwoven systems can leave us confused and embattled. For the more than half a century since World War Two, the paradigm and mindset of war have continued to dominate public conversation as well as scientific endeavors, and political argument too often falls back on war as the metaphor to address every challenge, major or minor, all in the archaic, mammalian-brain language of threats, assaults, victors, and vanquished. We see problems as conflicts to be resolved by force and dominance, battles to be won at all cost, rather than dynamic systems of flow to be better understood. We have waged the War on Poverty, the War on Drugs, the War on Cancer, the War on Terror – each of these declarations of war doing more to make its problems worse rather than better. War has defined the American mentality for all of the same decades that postwar industries have prospered from the technical inventions of war.

And, lest those who have not lost loved ones in the conflicts forget, the United States has been unceasingly at war since 2001 – Afghanistan, Iraq, North Africa, and the Middle East. And before this century began, the US spent the years from 1945 to the end of the 20th century involved in more wars than we might remember, or care to. Beyond the measurable losses directly from war, we have also suffered immeasurable losses on all sides from both the mentality and technologies of war. War is clumsy, always bringing death and destruction with its achievements, leaving too many of its warriors permanently scarred with intractable physical wounds and lasting mental disability. But driving these dreadful losses endured by many, there is always profit to be made by some – great fortunes to be enjoyed by the few.

And the costs of war are always borne by those who receive no compensating profit. As Major General Smedley Butler told us in 1935, between the world wars, 'War is a racket'. And because, among rackets, it is 'the only one in which the profits are reckoned in dollars and the losses in lives', the only way to end war, then, is 'to take the profit out of it'.

Instead, overseas wars have seeded their culture back home in the states, including the winner-take-all profit motive. We are now under siege from Big Pharma, Big Agribusiness, and Big Food corporations, and we have unwittingly cooperated with their assaults on our wellness, even as they have sacrificed American lives and health for their own short-term profits. We must rally to defend ourselves. The enemy is here at home, not only in these massive industries but also in the public agencies which enable them, allow them, and fail to protect us from them. If most people can agree that a primary role of government is the protecting of American lives, and if unsafe food and drugs are a major threat to those lives, then surely we can enlist broad support across the political spectrum against these profiteers.

Legacy of world war

Before Eisenhower's presidency, World War Two had urgently compelled academic science to restructure itself to accommodate the war effort. Institutions had swept aside the obstacles between them, and tradition had given way to an era of interdisciplinary collaboration across multiple fields. Collective enterprise linked together and united to target specific challenges with key projects, necessary if the war effort was to succeed. Such shared endeavors had required a national imperative of sufficient magnitude – world war – to unite political support, mobilize funding resources, dismantle barriers, and launch multidisciplinary collaborative ventures like the Manhattan Project.

In wartime, secrecy is critical, and during the war, only those who absolutely had to know were aware of initiatives such as 'the bomb'. These novel experiments also brought new dangers far greater than predicted, such as radiation exposure, and science generally found itself in uncharted territory. The risks were not only unmeasured but unknowable. The bombings of Hiroshima and Nagasaki finally ended the war in the Pacific, but they did not obliterate the formerly secret megatons of nuclear capacity nor their unknown dangers; not to be wasted, then, these stores were redirected towards megawatts of nuclear power production. Thus, just as synthetic nitrate found commercial use as fertilizer without changing its explosive character, nuclear fuels too would still hold their fearful capacity. A small elite of nuclear experts emerged from World War Two, and they closely guarded their new knowledge, which they continued to acquire throughout the Cold War years, considering it a matter of extreme importance to national security. Over time, however, other countries would pursue their own initiatives to advance atomic science, finally breaking open global markets to commercial opportunities for developing atomic energy.

As it was for the Manhattan Project and then nuclear power, so it was for many other projects that massive government investment had spawned during the war years – whether chemical, pharmaceutical, plastics, refrigeration, automotive, electrical engineering, or telecommunications. These technologies were built on trade secrets and protected by patents, and the select group who gained ownership of these patents became the prime movers and pioneers of their day. Today, none of their wartime technologies or skills have been lost; rather, they have been put to new uses, and many of our most familiar technologies trace their roots back to the 1940s. To consider one powerful example, radio and television messaging has morphed from weekly wartime newsreels and morale-boosting propaganda into 24-hour streaming news and non-stop commercial advertising – now, increasingly, showing the roots of its propaganda heritage. Some of these wartime projects helped to launch the most successful blockbuster industries of the 20th century, and their leaders would become the titans of commercial enterprise and financial success.

Eisenhower warned against just such 'acquisition of unwarranted influence' as has manifested today in the mammoth power of our corporations. These businesses were built with industrial innovators' energy and diligence during the war, but their sons and daughters inherited them unearned. Eisenhower warned that 'only an alert and knowledgeable citizenry' can ensure that industry uses the machineries of war towards 'peaceful goals and methods', and such a citizenry would necessarily include many perspectives: company directors, shareholders, and investors, as well as employees, teachers, customers, and the general voting public, all thinking long-term about future events, each with a part to play in shaping a worthwhile future. But the boards of directors that arrived to make key decisions were instead ready to exploit what had been built – having had no part in the sacrifices of war but ready as opportunistic beneficiaries of the new status quo: of unprecedented wealth and influence over whole sectors of the economy.

Traditional companies tend to be shortsighted in their pursuit of building profit, and contemporary companies, in the digital age, especially so. Making products, honing efficiencies, and promoting growth are worthy basic tenets for building a company. But we have learned by now that, collectively, to create sustainable businesses as part of a sustainable economy, any industry must also optimize its use of our shared resources: exercise care with our raw material supply, maximize efficient energy use, and minimize environmental impact such as pollution to our air and water and the soil that grows our food supply. Instead, over the last 70 years, mid-20th century mass industrial methods have come to dominate areas of the economy to which they are ultimately ill-suited. Agriculture is an especially poor match for employing unsustainable practices such as the overuse of synthetic nitrate fertilizer, GMO monoculture designed to support the excesses of Roundup, and other herbicides and insecticides that disrupt vital ecosystems.

Industrial goals and farming principles present a profound contrast and mismatch with one another – an asymmetry – the former driven by short-term profit to milk the maximum from the available resources at all cost, the latter requiring respect for the long-term needs of the land and for honoring sustainable methods, to survive. Best practices in farming necessarily mirror the natural ecosystem and maintain a cyclical balance, in which everything is related to everything else – the land is a shared resource and water runs downstream. Individual farmers, needing to protect their investment in a modest plot of land, necessarily live under Eisenhower's directive that we 'avoid the impulse to live only for today, plundering, for our own ease and convenience, the precious resources of tomorrow'. But large corporate agribusinesses, which control most US food production, can afford to plunder the precious resources in any one place, then abandon it for another – and they do. Environmental impact studies reveal innumerable examples of agribusiness's devastating impact, few more profound or sobering than the nutrient-rich runoff that feeds the harmful algal blooms in the Gulf of Mexico, their dead zone now larger than the state of New Jersey.

World War Two thus created many of the technologies, industries, and production capacities that define life in today's world, both its nature and its quality. In the span of 70 years, we have witnessed both phenomenal and phenomenally problematic changes in human health and in the planet's environment. Even as we have also seen less famine and food shortage, at least regionally, and as we have rejoiced in the virtual eradication of some threats to human health, including polio, we have seen the climb of preventable, slow-developing, and deadly threats to human health and life, through the very substances that are supposed to feed and heal us. It is time to adjust our perspective, to find a model that better fits with the realities before us.

While mounting our resistance to the assaults on health, we also know enough now to understand that the war metaphor, the war mentality, does not work for solving most problems, especially not over the long term, which is the view that symmetry demands we take. Symmetry-seeking asks us to look at what seems different and notice how it is the same. What do any two entities, which seem utterly dissimilar, actually have in common?

Symmetry also asks us to step back and take a still broader view, to consider the purpose and function of systems. Looking beyond any one business to these industries as a whole, to their own purposes, demands that we consider the vested interest of some of these corporations, in profit for their owners and stakeholders. There, in light of symmetry, we can see how the chickens are already coming home to roost, producing negative consequences up and down the flow sequence. Already, we see viral epidemics wiping out millions of turkeys and frequent recalls of tainted food products. Synthetic fertilizer runoff from farmers' fields has contaminated drinking water with excessive phosphorus. And nitrogen is the most widespread pollution problem of US waterways, as industrial farmers' short-term gains drive fisheries' long-term losses. Within this broad view, it becomes clear that only cooperation can succeed, and only metaphors of balance and sustainability – a paradigm based on symmetry and flow – make sense, because we are truly all in this together.

Information industries

Connecting the dots among the mega-industries that affect personal health requires knowledge from multiple sources, not just medical research. Complicating our knowledge-gathering is a foundational reality about our current economy: that it is built on garnering public attention through advertising in mass media. In the 1950s, the American economy was still based on manufacturing, but advertising and marketing exploded in the subsequent decades, followed by the digital revolution of the 70s and 80s. The resulting information age flooded us with our overload of data, including health data, a phenomenon inextricably interwoven with the information age brought on by computerization.

Now the commodity is not manufactured goods but knowledge, and profit-driven interests are the prime disseminators of information, proffered in the service of selling things of all kinds, even prescription drugs. Even more than information or knowledge itself, the prime commodity currently is attention itself: the ability to garner clicks and eyeballs and successfully persuade those eyeballs using whatever message one desires. Sellers present products in the best possible light, downplaying their dangers. And markets are presumed to be 'free' – with competition ostensibly holding prices within reason – while in reality monopoly and patent protection allow abusive pricing practices. Thus, all that remains standing between compassion and reckless profiteering is whatever ethical restraint drug makers or prescribers may choose to exercise, or not, even when the product is a lifesaving drug.

Fortunately, many of the same channels that distribute profit-driven advertising can also be used to fill gaps in knowledge and to disseminate correctives to misinformation, including truth that counters misleading advertising. Research has shown that such knowledge is empowering if it reaches the audiences who need it, even for adolescents buying soda. Greater transparency, giving accurate information about the available choices, enables an individual to select a healthier product, and many of us act on that knowledge. In a study of more than 4,500 adolescents purchasing sugary beverages, information about the sugar content significantly influenced their decisions not to purchase them.[1]

Reaching the American consumer creates upward pressure on suppliers to satisfy shifting demands. And reaching the American voter can create pressure on lawmakers to shape statutes responsive to citizens' needs. The wheels of legislation often turn slowly, but public attention to pharmaceutical companies' egregious abuses, such as the pricing of lifesaving Epi-pens, has prompted an outcry for laws to make such gross exploitation impossible. For now, however, such profiteering remains legal, and the campaign to bring public awareness to the pricing problem is arguably just beginning.

While consumerist thinking run amok is a blight on American culture and arguably our worst export, food is one product of which we are all literally consumers. We must eat to live, and when people change what they want to eat, the market must and eventually does respond. And it is information – knowledge – which creates that desire and demand, over time.

A full 70 years after World War Two has brought us into a world of new avenues for communication, far beyond the first-generation technologies the war originally spawned. We therefore need to make the most of these laptop computers, handheld devices, and the internet, towards ends that will genuinely improve our lives and health. The double-edged digital revolution brings a mix of blessings and curses: access to information, education, and entertainment, but also misinformation – malware, phishing scams, and invasions of viruses, uninvited cookies, and spam (each, notably, named after a disease or junk food) – including pornography, the junk food of intimacy. Amidst these downsides, the world wide web has greatly democratized knowledge, free to share and free to discover for anyone with an internet connection, and countless providers show altruistic interests towards health and wellbeing. Authentic sources can offer reliable knowledge to inform decisions or guide the hand in any number of inquiries, helping us make better choices (sometimes for free, sometimes for profit, and sometimes funded through bombardments of advertising). If we continue to leverage trustworthy resources and learn how to use them to greater effect, we can set the stage for us all being better informed and better decision-makers.

One of the greatest and still-growing assets of the digital era is 'Big Data' – some of it accessible through the internet, though much remains proprietary. At this moment, current technology has made it possible to collect – and is continually collecting – more data related to any imaginable subject than we know what to do with, literally. Our technological platforms are scrambling to develop the analytical tools to make better use of this stockpile of information. Algorithms and machine learning are one path to making sense of it all, and the methodology

of symmetry-seeking that we have been exploring in this book is another kind of guide to figuring out a path through all that data – towards cooperative flow as a means of understanding and then responding more wisely to what we now know, or can know.

Somewhat paradoxically, the first generation of all this data-collecting has been driven and funded by corporate marketing, by the desire to know ever more and more about consumers' habits of consumption, in order to better market products to them, including food and drugs. But if we look for opportunities to harness the data about health – insisting on access to information about broad public health trends as well as ownership of our own personal health data – we can apply symmetry-seeking to find patterns to help us change our eating and other habits, in turn pushing back on industry and on legislation, by asking for changes in response to our needs.

The great ship turns slowly. Remembering the 50 years it took to understand DNA and the 50 years or more to institute measures to prevent scurvy or cholera, we know that for some individuals, change will come too late. But for millions and billions of us, there is still time to turn the trajectory of personal health onto a new course, while mitigating damage to the planet as well, if we make better use of our systems for disseminating information and learning better how to evaluate and respond to it. In Chapter 9 we talked about personal decision-making and action, which is the single greatest direct influence any one of us can have on our health. Now, in keeping with symmetry-seeking principles and the interconnectedness of our biotensegrity architecture, we step back to see the broader view, of how our personal choices are linked to and affect the larger whole, as it, in turn, affects our personal health.

Collaborative intelligence

One person's efforts alone can feel woefully inadequate in the face of vast, multisystem problems, but if exercised in concert with others', one person's efforts can effect profound changes. Margaret Mead exhorted us, 'Never doubt that a small group of thoughtful, committed citizens can change the world' because 'indeed, it's the only thing that ever has'. Humans have consistently found ways to transform a community, a whole culture, and even the whole planet, over time. Sometimes we can even effect change quickly – especially as speed of change becomes a hallmark of our current globalized, technologized era. Fortunately, no one person need understand every aspect of the interlocking systems to play a part in shifting the direction of the whole, even in guiding or driving the shift.

Other species demonstrate the power of collective intelligence – including birds, whose brain hemispheres we met in the previous chapter. Notwithstanding their tiny bird brains, vast numbers of individual wild birds pool their individual inputs to common purpose, and one of their more extraordinary displays is annual migration. Each bird is equipped with its own internal compass and external sensors of time and space, ready to learn how to find a place in an array of their fellow birds and move the flock in the right direction. When the season arrives, great flocks gather together in the usual meeting place and wait for the stragglers to join, before rising in a cloud to follow the invisible meridian, at times sorting themselves into their wondrously symmetrical V-shaped formations. There is no appointed leader, no administrators nor officials to police the crowd: only an insistent internal compass built into the genetic code – inherited from ancestors who have successfully beaten the path with their wings, every year since migrations began.

Answering the inner imperative to move to new climes, each bird contributes to the movement of the whole by focusing its attention on its own spot in the line-up; the bird spaces itself between neighbors, pushing and nudging its influence this way or that, to find its own place. Every bird holds equal rank and status, the same weight of authority, as they move to different spots in the formation. The individual birds contribute their personal compass and bearings to influence the flight of the whole group. Each tiny bird brain has added to the collective whole and together, somehow (we have yet to fully answer how), they find consensus on timing and a flight path to their designated better grounds, either for feeding or for breeding.

The collective principle of flocking moves individuals into groups, within which higher orders of organization emerge, as each bird challenges the next and forces itself in place, jostling to balance and order the whole in a self-organizing pattern. Together the group will succeed where the individual might have failed. Nature loves this collective principle, grouping living cells into tissues and organs, then building organisms into groups, whether colonies of ants or communities of humans, always towards greater wholes and larger groupings. Whether through animal instinct or human consciousness, groups allow and require communication, collaboration, and competition – sharing the collective intelligence of the whole to make possible that which was not.

Humans, too, demonstrate how the group is more effective than an individual alone at assessing information, when they play a raffle game like 'guess how many pennies in a jar'. Ask different people and you will get widely different answers, some far too little and others far too large. But if you ask enough people and average all the hundreds of answers together, it is amazing how close the mean value will be to the true answer. This is the power of many, in a phenomenon first noted in Francis Galton's seminal 1907 article 'The Wisdom of Crowds', about collective success guessing the weight of a prize ox.[2] No one person really knows the answer, but the whole does, and it is the collaboration of collective intelligence that brings a reliable, reproducible, and dependable accuracy. (In his 2004 book of the same title, James Surowiecki helps to define the characteristics of a crowd most likely to outperform the experts – notably including diversity and independence.[3])

In addition to pooling our knowledge and our insight, humans also enact a more direct replication of birds' flocking behavior. Game theory predicts that in groups of individuals, it takes only a few of them to aggregate together to tip the balance and move the whole, whether spectators rising to their feet in a sports arena or a crowd moving to riot in the streets. Even in government, a few legislators voting consistently together, across issues, wins the day much more often than not. Such is the influence of coupling together into aggregates. Political rigidity, blindly following along party lines in preset factions and voting in caucus, reliably gums up the works and blocks the flow of rational legislative decision-making, and herd mentality can lend itself to mindlessness and destructive behavior. But mindful awareness, with each individual guided by a personal compass, can push or pull and influence the whole towards more constructive and meaningful decision-making.

Keeping expertise in perspective

Rather than relying on collaborative intelligence to solve our complex, interrelated problems, the current legislative and administrative models for managing our data overload place us in a predicament. Ultimate decision-making authority lies in the hands of the few – a handful of government officials, both elected and appointed, and a great many corporate executives. A better model for decision-making relies on the detailed knowledge from the many experts who study underlying issues, each immersed in a particular specialty topic. As Tom Nichols explains in *The Death of Expertise*,[4] it is experts who know the facts and hold the key information about critical parts. But they are not best suited to be policymakers, because experts' focus on detail can deprive them of the broad view, so that they miss the interrelatedness of the multifaceted whole. The biotensegrity architecture of our modern world means that change in one place brings changes throughout and consequences for all, so decision-making is necessarily very complex and needs input from experts across multiple fields.

Our lawmakers rely in part on experts who are paid not to be objective but to be lobbyists: to use information selectively, for influencing the decision-makers at all levels of government. Chemical

agribusiness, Big Pharma, and Big Food have effectively limitless resources for this endeavor, so they have assumed ever-greater control over decision-making – not a system that lends itself well to effective, far-reaching analysis. Competing ideas and narrow agendas lobby for support, promoted by advertising, over-simplified in the media, and polarized by political divisions that deepen rather than ease by the next round of elections.

In contrast, the intelligence of flocking distributes decision-making among truly independent individuals, and the collaboration works best when each brings knowledge and has some power to affect the whole. For our flocking abilities to work towards better personal health and sustainable systems, we must each take responsibility for the path of flight, thereby nudging our neighbors and in turn the full group in best directions. Knowing that our personal choices have an effect on the whole, on the group, on humankind, how then do we evaluate that overall direction, so that we can make better choices? How do we assess not only what food to buy on our next trip to the market but what institutions to support, what candidates to vote for, what companies to work for? Amidst the hubbub, where should we give our attention? The principles of symmetry-seeking are a way to approach these questions, to evaluate our overall direction, not only in our private lives but in our public engagement. You can assess your own expertise as you continue to make your own place in the flock, starting by comparing the areas you personally know best to any of the models we've explored in this book. How does a system you know well compare to a farm, or a cell, or a human body?

Eisenhower warned of the challenge and necessity of letting neither the leaders of the military-industrial complex nor academics dominate the other group nor the rest of the citizenry. We must be wary of the role of experts: not to ignore the value of their input, but to keep their guidance in perspective, by remembering a recommendation's source and its purveyor's motivations. Just as doctors are experts in disease but not wellness, the dairy industry experts know how to increase milk production but not so much about nutrition or the long-term health effects of dairy products. And for an old guard of experts, moments of discovery and phases of change are often difficult or confusing, even if the innovations are also exciting and make life easier for nearly everyone. The candle expert did not devise the gas lamp, any more than the gas expert developed the fluorescent light bulb. Imagination opens new paths to advance the human condition, and collective intelligence sometimes finds a novel idea promising, gathers around it, and shares it, providing a larger group the opportunity to vault past current limitations and onwards to a better paradigm.

In the years after World War Two, experts from many fields were given enough space to collaborate and combine intelligence, so that they could reach beyond former limits to create a new reality, and the 1969 moon landing exemplified the power of their cooperation. Unified by a common task, focused on the physical, logistical, and engineering challenges, NASA had a shared vision, shared obstacles to overcome, and a shared fate. Just as in the forward flow of a migrating flock, the teams of experts working together synergistically were able to find the energy and knowledge to forge an effective flight path – and to achieve much more than the individuals could have, if working in competition. More recently, medicine and public health gave us an extraordinary example of cross-disciplinary collaborative success: identification of the SARS virus. Funded by the World Health Organization (WHO), research teams from eleven laboratories in different countries collaborated on a global response to the alarming threat of an unknown disease epidemic first recognized in China. Redirecting all their efforts to work on the urgent challenge, they identified the responsible coronavirus within a few weeks and established an effective plan for treating it. If the collaborative intelligence of flocking is to help us, we need to learn how to bring the experts together, to move beyond limited views and limited understanding. Symmetry-seeking habits of thought can help to carry us beyond those limitations.

Teaming up for better health

Human health and wellness provide a perfect opportunity to put collaborative intelligence to work for all of us. Health is a shared concern, and illness affects every person and family at one time or another. Furthering the universal benefits of wellness and shouldering the shared burden of disease, all communities and organizations bear the costs of lost productivity, and multiple entities carry the expenses of medical care, including through shared medical insurance plans, through tax breaks that transfer costs to other taxpayers, and through taxes (in the US, Medicare and Medicaid). All of us have a vested interest in one another's health, but within the communities of health care providers, any possibility of collaboration is upended by commercial interest – competition or antagonism. The market forces that dominate the health professions drive turf wars between traditional medicine and less conventional strategies, which western medicine seems committed to opposing. But stopping the flow and interchange of ideas keeps out treatments, even those oriented to wellness and prevention, which provide a vital complement to medicine's focus on treating disease.

Diverse cultures and different traditions adopt highly disparate approaches to problems of health and wellness. What we call western medicine is unmatched in technological sophistication, diagnostic accuracy, and life-saving resuscitation, but its specialty-based treatment can be over-focused on isolated organ systems or disease-specific strategies, rather than whole-patient care. Conventional medicine emphasizes physical body parts, with much less regard for mental and emotional health, let alone the spiritual, familial, and cultural needs of the patient. Hospital medicine often pushes too hard to sustain life at all cost, without respect or recognition that death is as much an obligatory part of life as birth.

Normal life events like pregnancy and childbirth offer an excellent example of healthy milestones well-served by alternative approaches to care, as they are conditions within wellness, not episodes of disease. But western health care offers only hospital care and advocates a medicalized birthing process, even though some mothers express a strong preference for natural childbirth, water births, or hypnobirthing. Attentive monitoring and preventive care at intervals throughout the pregnancy are critical not only to prepare for delivery but also to recognize, avoid, or manage potential problems. Many of the risk factors are known well before childbirth, which is after all the crowning event of months of pregnancy.

Some of those risks for mother and child are already set by the initial conditions, such as the age of the mother to be – for instance, adolescent girls' babies are at much greater risk during pregnancy and childbirth than the babies of mothers who themselves are fully grown. And a mother's decisions and actions regarding her health before and during the pregnancy lock in much of the trajectory for better or worse health outcomes well before she reaches labor. With no known extraordinary risks, many mothers-to-be challenge the prevailing dominance of Big Medicine by opting for a midwife or doula.

Like other health care needs, obstetrical care is both art and science, personal and general, homely and institutional: it can avail itself of many methods of practice, still balanced with state-of-the-science technical facilities and resources. And even though delivery is a normal life event, the human pelvic floor and birth canal are not always well-suited to accommodate the birthing baby's large head, as we discovered in Chapter 2, so some mothers need special methods of expedited delivery such as cesarean section to protect or save the lives and health of mother and child. Acknowledging this risk, an ideal setting provides full resources for a healthy, non-medical birth while also working in close cooperation with full medical, surgical, and intensive-care services, their capabilities readily available in the event of an obstetrical emergency. What stops such collaboration is a lack of flow between methods of care and bodies of knowledge.

Like the choices for natural childbirth, some patients are drawn to acupuncture, osteopathy, chiropractic, Ayurveda, physical practices, naturopathic methods, or energy work. Doctors trained in the western tradition, towards a narrow focus, often resist cooperation with such practices, ostensibly for the health of their patients, assuming alternative therapies to be unproven or perhaps even dangerous. But in the current structure of the medical marketplace, physicians have a deeply vested interest in ignoring or impugning non-medical approaches to health, ensuring that those methods remain untested and then, necessarily, unendorsed by the medical establishment, because those alternative therapies pull patient dollars away from Big Medicine.

Whether conventional or not, most methods of care share fundamental features: taking a history, physical examination, differential diagnosis, and strategies for treatment. And widely varying modalities of care have genuine protagonists with broad clinical experience and a real sense of what value they bring to their patients. Each therapeutic approach boasts loyal clients who have gained genuine personal benefits from its treatment programs. So while of course it makes sense for laws and regulations to protect citizens from truly harmful practices and useless methodologies, the vested interests of practitioners in the medical marketplace, who have already gained certification to treat, tend to fiercely resist encroachment from competing interests. Newcomers face stiff resistance, roadblocks obstructing the flow of health and wellness practices. If alternative strategies succeed, they challenge the disease-oriented paradigm and perhaps even threaten to undermine some of the money-making side of the scientific enterprise.

Sharing the medical marketplace

In any case, data so far suggest that the alternatives to pills-and-procedures medical care are not replacing medicine but are being employed alongside it – that is, complementing medicine, not substituting for it. In a US survey, approximately 38% of adults and 12% of children reported using complementary or alternative methods, and a systematic literature review confirmed that a considerable proportion of the population include these strategies in their health care.[5] So while some in the medical establishment express concern and alarm about the growing strength of complementary and alternative strategies, the survey report estimated that 83 million adults spent $33.9 billion on such therapies in their efforts towards achieving and maintaining wellness, accounting for 11.2% of total out-of-pocket health care expenditures.

In some quarters, traditional medicine has already begun to embrace the benefits of complementary strategies into integrative medicine, including those with roots in ancient eastern traditions. For instance, Jon Kabat-Zinn, who has spent his career on the faculty of a mainstream medical school, adapted what he had learned about spiritual meditation techniques to develop and share his mindfulness-based stress reduction program – and to scientifically test their effects on health. A randomized, controlled study of meditation therapy for African Americans, who have a high rate of cardiovascular disease, followed 201 patients with coronary artery disease for more than 5 years and found a 48% risk reduction in death, heart attacks, and stroke for the group that used meditation.[6]

The problem most frequently treated with complementary therapies is pain, including back pain, neck pain, and arthritis. Pain is hard to treat because it is not simply in body parts but affects one's whole being. Pain can persist long after its initial triggering event is over, and it can move about the body in all sorts of squirrelly ways, even taking refuge in a phantom limb. Also, as discussed in Chapter 9, certain aspects of early life such as secure attachment and lack of adverse childhood events have a role in shaping resilience, and for those with a less favorable history, the body can hold an imprint, a memory, of past experiences, including pain. Such experience can then shape the present awareness, such that two people with a similar condition can experience pain

in very different ways, making it unpredictable and particularly hard to treat.

About 100 million American adults suffer from chronic pain, according to the Institute of Medicine – at an estimated cost of $560 to $635 billion in healthcare and lost productivity.[7] By these measures, pain management would be a more expensive problem than heart disease, cancer, or diabetes. But the pills and procedures that these figures assume as the default treatments are not always the answer for pain. Even powerful opioid prescription drugs become much less effective over time, and pain medications carry their own risks, ranging from drug interactions to addiction and fatal overdoses, and so doctors try to avoid prescribing them for long-term use. Still, in spite of national attention, intense medical awareness of risk, and much talk about reducing opioid use, prescription drug overdose is the number one cause of accidental death in the US.[8,9] Clearly, we are in need of major shift in our approach to managing chronic pain.

Many pain sufferers are drawn to complementary strategies such as acupuncture, but on the whole, physicians hold out against such alternatives. Both following and leading physicians, the payers and insurance companies typically resist or refuse to bear the costs. And even when doctors' own patients report positive treatment outcomes from complementary therapies, they may remain nonplussed about any method not tested within their own research tradition, so that they cannot understand nor explain how it might work.

Still, despite the obvious asymmetry of power favoring the established medical professions in the present models that control the business of health care, it would be fair to say that a grudging acceptance of non-traditional complementary strategies is leading to more integration, respect, and collaboration between different views of health and healing. And surely the current opioid crisis points to pain management as a compelling opportunity to move away from pharmaceuticals and develop or expand complementary services. With customers as the most powerful constituents, patients ultimately will have the power and ability to reshape the medical marketplace, in proportion to the openness of the medical market and the quality of accurate information about its offerings.

Wellness is not for doctors to manage

In the meantime, we know from the history of medicine – its successes and failures at tracing and treating disease, as we saw in Chapter 8 – that medical research can be a slow-moving ship, and that doctoring is not often the place to look for preventive health measures, such as those we need for addressing the underlying causes that lead to pain, some of which are also connected to the obesity epidemic. Nor is medical care the best route to providing health maintenance or supporting wellness initiatives. I know too well, because I'm a doctor interested in wellness and prevention, and I know many providers of conventional medical care who want only health and wholeness for their patients, but we don't see them until after something has broken. (For prevention, our best hope would be to make sure family members are well-informed about risks and strategies, as they may still be enjoying good health but carry similar vulnerabilities.) Only when outcomes manifest as dramatic, life-threatening diseases with numerous victims does medicine rise to its most impressive, finding a reliable method to interrupt the crisis for individual patients and save lives or limbs, redirecting medical warfare on a known enemy to health and opening fire on the attacker.

And while modern, western medicine may not be the best system for delivering wellness, it does offer a handful of outstanding models of collaborative intelligence. Our best tertiary medical centers – those providing highly specialized or extended care, well-funded and affiliated with medical schools – function as brain centers rather than bed centers, which is why patients with the awareness, resources, and access to such hospitals seek them out when they face a dangerous or knotty health problem. Medical training involves teaching and learning, presenting and discussing cases, common and uncommon diagnoses, and such discussions include all relevant

organ systems, associated conditions, and disorders. In such an environment, doctors and doctors-in-training, nurses and nurses-in-training are always balancing risks and benefits. They consider a wide range of factors, evaluate multiple opinions, and create a collaborative consensus for individual care, as symmetry would advise. Teachers are immersed in the learning environment and teaching is learning, so questions from students and colleagues inspire a quest to search again for better answers, driving research to provide a deeper understanding of disease.

In all major medical centers, then, patient care, teaching, and research coexist as a complementary triad, and the best include all of the health care team in the process of pooling knowledge: nurses, technicians, and administrators. Most importantly, the best care involves the patient and the family, because medical care is always better when the patient and caregivers are well-informed and active collaborators in caring for their own health. A knowledgeable patient and family can and should be partners in care, empowered to take an active role in the collaborative decision-making.

Sometimes, eventually, medicine may also look upstream to causes or helps to collect data, which can become valuable knowledge in other hands, for prevention. Our institutions of public health especially, such as the Centers for Disease Control, collect and analyze just such information, using just the sort of principles we understand here as symmetry-seeking, to make sense of inputs from a full range of disciplines and types of expertise. With enough support, such institutions will prove critical to the 21st century task of facing the mountain of data and illuminating the patterns that affect our health. With that kind of knowledge, health practitioners of all kinds and individuals themselves will be increasingly empowered to make better health decisions for themselves.

Managing chronic illness

Symmetry-seeking asks us to look upstream and downstream, comparing to similar occurrences, systems, or events. Once a health problem has already arisen and we are evaluating how to approach it and whose help we might seek, we should consider how far downstream we have come already. Acute-onset, severe, and life-threatening problems warrant rapid and sometimes drastic medical or surgical measures. These are the times where western medicine is at its very best, offering high-tech intensive care and unmatched success rates. When looking for the pattern most likely to benefit from heavy medical intervention, we can consider a spectrum of severity and urgency. An appropriate response to the life-threatening health crisis of kidney failure, for example, may be dialysis or organ transplantation, saving a life and restoring a fully vigorous, healthy state. We know that end-stage organ failure is life threatening and anything less than heavy medical intervention will be useless, at which point we must accept the greater cost and the increased complexity of managing the problem, including a much greater risk of possible complications from the treatment.

On the other hand, most disease is not severe. Even a mild disorder is annoying, of course, and others are worse, sometimes even disabling. But most are not life-threatening. The most common kind of health problems are the chronic conditions, impacting quality of life but not the possibility of life itself. Many of these problems will be improved by complementary strategies that are much less drastic, less costly, and less risky than common medical treatments, often with professional intervention but not from doctors. For example, chronic persisting pain in the knee has been shown to improve with physical therapy, just as much and as often as with expensive state of the art arthroscopic knee surgeries.[10]

When seeking treatment for knee pain, patients know when they have gained benefit, if their knee stops hurting in response to treatment. In such cases, patients allowed access to a marketplace of treatments are good judges of outcomes. They can recognize when other therapies succeed or fail as well, such as osteopathy, acupuncture, or physical manipulation. And practitioners themselves, too, learn from their patients what works and responds

best to their practice patterns and recommendations. Reinforced by clinical successes they observe in their practice, most are convinced of the genuine benefits of their favored recommendations, including the nutrition specialists who might peddle products with their advice.

Health care practitioners tend to align with one camp or another, in competition with the other tribes treating ill health and promoting wellness. But if they set themselves in conflict with other approaches to health rather than confluence, the different schools often know very little, if anything, about one another, paying little attention to experts outside their own field or data collected and summarized for different types of practitioners. And for consumers to receive the best possible support for optimal health, they need access to the full range of resources and experts, who in turn allow information to flow freely between them, maximizing their understanding of the human body and including their fellow experts, in widely differing fields, to complete the picture. Each specialty holds knowledge and understanding about some part of a whole, but optimal healthcare requires an all-inclusive view that includes the whole.

Cooperation and collaboration across fields and disciplines is especially important when it comes to prevention. Where crisis intervention for catastrophic health events is all about speed, maintaining good health is slow, invisible work. While we may experience some immediate or short-term payoff in energy, strength, or optimism from our slow drops of investment in health, substantive success or failure may not reveal itself until years later, having been masked by the absence of various ailments during middle age, or by our good walking ability into our nineties.

The best health care isn't a treatment, because it doesn't involve a problem – there are no symptoms to bring one into an office. So, to design the best possible programs of prevention and health maintenance, we need the best possible data and analysis from all possible fields: upstream from the diseases doctors treat, to avoidable causes, and downstream from all manner of practitioners' treatments, to find out what works – which sometimes works too as prevention, such as the diet prescribed for a diabetic that might have prevented the diabetes in the first place.

We need complementary medicine in an open flow conversation with traditional medicine and both in conversation with the analysts of public health. Professionals and researchers in any of these fields need to stay alert to relevant information from the others, as well as to information from other fields such as agriculture, the environment, nutrition and food production. And as literal consumers not only of health care services but of food, we can stay alert for joint statements from multiple professionals together, such as from multidisciplinary commissions – or when different types of practitioners give similar guidance – that tell us when their data align. And health and wellness professionals can learn to pay special attention to multidisciplinary information, as we seek to guide patients with the best possible health recommendations.

As our individual lives flow from youth to old age, wellness drifts a bit over the years, perhaps revealing some hidden deficits such as we saw in Chapter 2. Our full range of agilities and abilities slowly lessens, and silent limits begin to constrain our experience. Fortunately, we can become increasingly attentive to preventive health over those same years, in plenty of time to make a positive impact on our long-term trajectory. Responsibility falls to each of us to try to learn as much as we can about how to nurture our own wellbeing, with attention to diet and exercise – to do the work of keeping the peace in our bodies. Fortunately, our personal efforts simultaneously influence the flock to improve prospects for future generations as well, with our choices in the medical marketplace like votes for the future. The data tell us that we can measurably reduce the likelihood that the big guns of drugs, surgery, and hospitals will need to come in and wage war on some part of bodies to save us from the brink of catastrophe. We can improve our range of function and even extend our years of productive living.

Both providers and consumers of treatments and wellness care have an opportunity to shift the way we approach the tending of our bodies and minds. Moving forward is about collaborative exchange and adjustment for the whole, integrating and including all in a process led by imagination, integrity and synthesis. And within that framework, the individual must begin to assume a much stronger personal role. As we become better informed and more engaged, we cannot only help to shape health care services but become more engaged decision-makers in our own health.

Time for change

Not only in medicine but in all branches of science, competing claims and hypotheses wrestle at every step to make their just assertions, protagonists and antagonists challenged to prove or disprove each other's claims. Such is the nature and method of science. But Thomas Kuhn taught us that the history of science has not been of steady progress towards a shared goal; rather, it has been marked by intervals of quiet acceptance of a status quo, punctuated by episodes of major revision and restructuring. Science collects new data all the time, some adding to the fit and reinforcing existing theory, but other evidence conflicting with it. Over time, a weight of testimony accumulates, until the balance of belief reaches a decisive tipping point and cannot withstand the overwhelming pressure of new facts, and then the old paradigm must collapse and a new model take its place.

Our paradigm of caring for human health is overdue for a major shift, as the status quo has not changed since 1940s. The tools of symmetry-seeking ask us to look for methods and systems that can teach us a better way, and to consider approaches and structures beyond what we already know. The current methods of attempting to care for human health, almost all funneled through the medical establishment, as it is, are clearly facing overwhelming pressures, and we need new methods, new systems, and a new paradigm that preserves the genius of medicine while going beyond it to integrate health – using a global approach that honors the best of varied traditions and international know-how while at the same time making full use of science, recognizing that our best solutions lie at frontiers currently just beyond our reach. We know a great deal, and we need new ways to apply what we know towards greater, all-inclusive human health.

The broadest possible perspective understands human health as one system in relationship to many, many others – not at war, as competing systems, but in careful communication that opens the flow of knowledge. To say that we are all in this together is to acknowledge that it is not enough to interest ourselves in parts, no longer acceptable to consider only the immediate needs of today and the next few years, when we need a much broader and more inclusive, long-range view of the whole. Despite experts' contributing so much knowledge and detailed understanding, we have yet to act wisely on what we've learned, as specialization has effectively blocked comprehensive thinking. The span of years since World War Two is similar to the course of just one human lifetime, and we now see many signs emerging of the unsustainable practices that threaten future prospects and our personal health.

Fortunately, advances in communication technology hold possibilities, also new in human history, for maximizing collaborative intelligence, if we can sort out how to make best use of experts amidst the cacophony of vested interests.[11] The global reach of the internet has begun to offer mechanisms both for disseminating information and for increasing our collective wisdom. Free access to shared space such as wikis, discussions on blogs, and crowdsourced data collection invite participants to create sites of sharing. There we can pool and analyze cutting-edge knowledge to challenge and revise assumptions, improve resources, and create a landscape of learning that echoes organic diversity and growth, drawing in more and more of the migrating flock to mold and be molded by shared knowledge and collaborative intelligence. And, used wisely, these sites of information-sharing also provide virtual meeting places where participants can coordinate collective action.

Gradually, we are finding our bearings in this new landscape, learning to distinguish between trustworthy sources, wild ideas, and paid advertisements. Even in a setting of imperfect or incomplete data, in the spaces where the science is not yet decided and the answers aren't obvious, the collaborative approach and symmetry-seeking tools can be helpful in deciding which scientific knowledge is most relevant. Looking for balance, we can move forward cautiously towards our shared goals of better health and wellness.

References

1. Sara N. Bleich SN, Barry CL, Gary-Webb TL, Herring BJ (2014) Reducing Sugar-Sweetened Beverage Consumption by Providing Caloric Information: How Black Adolescents Alter Their Purchases and Whether the Effects Persist. American Journal of Public Health 104(12): 2417-2424.

2. Galton F (1907). The Wisdom of Crowds. Nature, 1949 (75): 450-451. Online: https://www.scribd.com/document/19769105/The-Wisdom-of-Crowds.

3. Surowiecki J (2004). The Wisdom of Crowds. New York: Little, Brown.

4. Nichols T (2016). The Death of Expertise. New York: OUP USA.

5. National Center for Complementary and Integrative Health. The Use of Complementary and Alternative Medicine in the United States: Cost Data. Online: https://nccih.nih.gov/news/camstats/costs/costdatafs.htm.

6. Schneider RH, Grim CE, Rainforth MV (2012). Stress Reduction in the Secondary Prevention of Cardiovascular Disease. Circulation: Cardiovascular Quality and Outcomes, 5: 750-758.

7. Pizzo PA, Clark NM (2012). Alleviating Suffering 101 – Pain Relief in the United States. New England Journal of Medicine, 366: 197-199.

8. Katz J (2017). Drug deaths in America are rising faster than ever. Online: https://www.nytimes.com/interactive/2017/06/05/upshot/opioid-epidemic-drug-overdose-deaths-are-rising-faster-than-ever.html.

9. American Society of Addiction Medicine (2016). Opioid Addiction 2016 Facts & Figures. Online: https://www.asam.org/docs/default-source/advocacy/opioid-addiction-disease-facts-figures.pdf.

10. Katz JN, Brophy RH, Chaisson CE et al (2013) Surgery versus Physical Therapy for a Meniscal Tear and Osteoarthritis. New England Journal of Medicine. Online: https://www.sciencedaily.com/releases/2013/03/130321133244.htm.

11. Shorenstein Center on Media, Politics and Public Policy (2017). Information Disorder: An interdisciplinary framework. Online: https://firstdraftnews.org/coe-report/.

Conclusion: using the symmetry-seeking toolkit

In *Seeking Symmetry*, we began on a human scale with 500-year old drawings of Leonardo da Vinci and the 100-year old physical laws of growth and form articulated by D'Arcy Thompson. In the present moment, the biotensegrity concepts of Stephen Levin explain our architecture: not only the workings of bones, joints and movement patterns, but also the development of the human embryo and fetus. The healthy human body is not the sum of its parts; it is an integrated, balanced whole in which every part is connected and dependent on every other. For the future, we must insist on a wider all-inclusive view of wholes.

Our brains see the world in pictures. When we encounter complexity, we seek out patterns and symmetries within it; once we have recognized a pattern, it becomes a concise form of storage for complex information. This smaller, more manageable packet can contain the larger whole, replacing the panoply of details with a memorable, visual essence of the parts' sum. Finding similarities makes complicated knowledge less difficult to grasp. Seeking symmetry also joins dots, showing how one kind of problem resembles another and thus helping us see how a solution to one problem might offer a better way out of another. Ultimately, pattern reveals structural explanations that link problems and ideas, connect one function with the next, and help to predict how the whole might work as one.

Mindful engagement

From multiple, independent disciplines and traditions, a contemporary approach to health has been emerging, one focused on nutrition, exercise, bodywork, energy, and wellness strategies. A common feature of this approach is a bigger-picture perspective that links differentiated elements and connects separate parts of systems to one another – between systems in the body, between the body and mind, and in interpersonal therapies – implicitly seeking what is similar, not just what is different.

The principles of symmetry-seeking explored and applied throughout this book are a particular approach to that big picture, grounded in questions of evolution and development:

- looking at the integrated whole, not just parts
- shifting perspective
- adjusting scale
- examining function and adaptive purpose
- looking upstream, back to a system's development
- looking downstream, forward to the long view.

We follow these principles with an eye always to balance and imbalance, symmetry and missed symmetry, and a system's tendency to compensate for disproportions, as it tries to right the ship back to equilibrium. We keep in mind that an integrated whole tends to invoke a self-regulating property that creates harmony and balance. As we process information, these are themes to look for, to make sense of existing research and to find a place for new knowledge as it emerges.

To make use of the tools of symmetry-seeking, we must first quell the fears that trigger poor judgment – the reptilian or mammalian responses that deny our prefrontal cortex and render it unable to do its work. Unfortunately, multiple public entities have a vested interest in provoking those primordial triggers and hormonal cascades that displace thoughtful behavior. Anger-making, fear-mongering, and crisis mode prove perennially useful for anyone wanting to provoke a reflexive response, including marketers trying to drive a rush to impulse buying and media outlets seeking to drive up viewership and clicks, but such reactivity will not help us evaluate data when analysis is needed. When threatened, we must apply what we are learning about the brain, with strategies such as pause-and-plan mindfulness for managing and deflecting default animal reactivity. We will need some serious participation from both our cerebral cortex and our prefrontal cortex – left brain for details, facts, and figures and right brain for the broader worldview, ethics, creativity, and contemplation – as we reflect and analyze.

With a calm mind, our first task for evaluating health information is usually to decide what is true, as discussed in Chapter 9. And the systems perspective that reveals our necessarily social context prompts us to remember a key factor in assessing the credibility of an information source: follow the money. What vested interests are involved in the chain of information delivery? Symmetry-seeking prompts us to look upstream as we consider the purveyors of the data – how are they funded, and what is their purpose? And we must look downstream: who stands to benefit from the likely responses to the information at hand? A money trail in either direction that leads to a dead end, with unknown benefactors or beneficiaries, is a cause for concern. And a party's having a vested interest in an outcome does not mean that its information is necessarily untrue; sometimes those with most to gain from a discovery also understand it best and present it truthfully. We simply should proceed with caution when considering their input, always looking for confirmation from other sources.

Act on facts

As this book has explored in depth, today's elephants in the room are the diseases of food. Heart disease, obesity, diabetes, and cancer are just some of the diseases with a documented rise in risk from certain well-known eating habits. What we eat is the one factor within our control that is most likely to determine our wellbeing. And, even with some prevailing unknowns about our food supply, the science is absolutely clear that the food you eat is the strongest predictor of your current and future health.

It is empowering to realize that each of us has authority to control our own health to a significant degree, and symmetry offers a missing element: the structural explanation of how health works in the body. Medicine brings valuable assets in diagnosis and treating diseases, but staying healthy, preventing and avoiding disease, is not up to doctors: it's up to us. It is no longer appropriate for doctors to be the sole directors of medical decision-making, claiming that the body is just too complicated for patients to understand. Armed with a structural explanation, patients and families can grasp the essence of a problem and become active partners in decision-making. With personal behavior as the most potent predictor of future health and wellness, today is the best day to start.

The first, repeated, and most important step towards health happens on your next trip the grocery store or farmer's market, when you select more fresh vegetables, fruits, and whole grains and less highly processed or nutrition-poor junk. Developing the habits of buying and eating healthful foods and reducing your intake of junk can be a challenge for many of us, if we are not in the habit already. But there is an abundance of reliable information to support us in the effort of building healthy eating habits too, if we have the patience to find, evaluate, and implement the guidance, some of which includes the elements of positive social engagement or intermediate rewards that we looked at in Chapter 9. And simply learning more about which foods are safer or healthier can carry us a long way towards better eating, by helping us understand what to eat and why we would want to choose certain foods over others. If we engage in finding and exchanging our knowledge of nutrition, of recipes for healthful meals that taste good, and of food safety, we participate in the collaborative enterprise of moving all of us towards better health and longevity.

Changing the buying habits that support your eating habits will not only affect your personal health but set several other processes in motion, including market pressure on food producers to grow and market more healthful food options. We see the effects of these efforts already, with the growing presence of health-oriented major grocery chains, a proliferation of local farmers' markets, and the growing popularity of farm-to-table restaurants. Better still, we have seen the 'health food' sections in ordinary grocery stores expand and efforts at better nutrition (and improved nutritional information) on most aisles of the supermarket.

In choosing what to eat, we can concentrate on buying only foods that we actually know what they are. Any item not from the produce section – those bought in a box, bag, can, or carton – should have a thorough ingredient list and detailed nutritional information, and the best foods will not contain unrecognizable ingredients, which usually are simply not foods but non-food chemicals. Best fresh foods, including produce and meats, poultry, and fish, will contain key information about how the product was raised, and what food fed your food. If vital information is missing, ask your grocer about the product's source and request from producers that they include the information in future labeling – easy to do in the age of email and social media.

Efforts to buy better food for yourself nudge the flock in a healthier direction – not least your loved ones, since research shows that the health habits of those closest to us strongly affect our own. And your little changes will not go unnoticed in Big Data. Big Food manufacturers, distributers, and industrial agribusinesses constantly monitor product sales, and they use promotions, price adjustments, and special marketing like the shepherd uses his dogs. With a bit of pause-and-plan, the well-informed consumer has

the power to shape the marketplace and collectively decide what will be more readily available for all of us to eat in the coming months and years.

Beyond our all-important individual food choices, the biggest push we need to make is insisting on knowing what we are eating. This information does not even require Big Data: simply basic information. A cooperative society with an infinitely specialized division of labor means that we are not each going to go back to growing all our own food – there would be time for little else, as past generations could attest. So, as we must eat food produced by others, we must press them to publish the simple facts about their products: what are the ingredients and how was it produced, processed, and packaged? Real food is easy to label, as it has few ingredients – often only one – but labeling for processed food must be written in comprehensible language, including information about additives and risk factors.

For animal products, we need to know about antibiotic and hormone use, types of feed, and containment; in plant products, pesticide and insecticide use and genetic modification; in packaged foods, the balance of ingredient types; and in all foods, clear language about synthetic chemicals used in any part of the process, with a description of their risks. Some of this information is already a part of labeling, but some of it still needs to be added, with its presentation simplified for consumers' clear understanding. Experience suggests this update will be a long slow process, because Big Food, like Big Tobacco, is likely to do everything possible to resist labeling standards that might reveal truths that threaten sales of their disease-promoting products. Big Pharma already has to list the risks associated with pharmaceuticals, in exhaustive detail, and symmetry-seeking demands that Big Food do the same.

The bigger picture: policies and people

If you want to go further in engaging on the question of food and health, beyond the need for full disclosure on labels, you can also begin to learn more about our food supply. Ask questions about a food's safety. Look upstream and downstream at production, to track food raised here and elsewhere, to understand the issues affecting the food itself and other systems, such as farming practices or water quality. Then you will be prepared to support local or national policies, policy-makers, or the businesses that insist on best practices, employing sustainable methods and producing safe, nutrient-rich foods.

When choosing what food practices to support, our priorities need to be transparency and wholeness – seeing the whole reality about and between food-production industries. For now, the status quo guides policy, because the current winners have the loudest voices, with leaders who have already acquired the resources and influence to shape the debate and to exclude alternative players. Established winners choose their experts, setting the language and the rules that govern current conditions and hold those edicts in place. When we consider that some of our leaders' decisions are allowing considerable harm to health and our planet, it is worthwhile to assess how best to engage with them for change. It will take a shift in viewpoint, such as symmetry-seeking provides, for them to see these shared problems in new ways.

Looking for similarity rather than difference, it is useful to remember the human scale: that the people who make decisions across these various industries and agencies – who work in the pharmaceutical industry, in industrial chemical agribusiness across America, in hospitals, in food manufacturing, and in the government – are facing the same challenges as the rest of us. They see their own loved ones with cancer and heart disease, and they face tough choices at work balancing short-term needs against long-term, as well as concerns about what they buy at the grocery store to feed themselves and their families. This shared experience gives us a starting point for seeing them not as enemies but as potential allies, if we can make use of the symmetry paradigm – rather than a war mentality – for finding common ground, for connecting the dots between their experience and others', for shifting their viewpoints toward more sustainable practices. An expansive view of the problems and possible solutions, whether for health or the environment, includes those shared needs and objectives that we have in common and within the whole, balancing between systems.

When we are evaluating the accuracy of new sources, we also need to learn how to remember whatever we have already learned. We can then see further upstream and all the way downstream at how various stakeholders' interests are vested in a system, affecting who is open to change and how to approach and engage them.

While fast-arising innovations in Big Data will be invaluable to forward motion, some of the most important contributors to health and to sustainability are simple and low-tech, as much about what we don't do as about what we do or consume. Such strategies are the least likely to be major money-makers for anyone, while some of the forces that work against health are wildly lucrative. Getting a good night's sleep or taking a walk at lunchtime are not likely to make many millionaires (a few, perhaps, with apps and supplies to help you do either). Instead, a screen full of ads on a gadget that keeps you awake past your bedtime and a restaurant plate piled with enough non-nutritious calories for several people's lunches are a much more direct route to others' wealth. But we need to find our systems, social supports, and methods for ensuring that we make these personal investments in health anyway – to keep finding and sharing crucial knowledge, and to promote institutional support and funding for sustainable practices.

And when some who can protect our food supply or our environment show themselves to be disingenuous mercenaries, or when the profiteers insist that their only responsibilities are short-term, to stock prices, we must attempt to show them how they are failing even that constituency, as their own returns diminish in the face of unsustainable practices. Problems with monoculture, super-weeds resistant to herbicides, super-pests resistant to insecticides, soil erosion, and soil depletion have already begun to create problems with producers' own yields and harvest, affecting sales and shareholder value. Poor agricultural practices are also hurting the bottom line of neighboring farms – surely a case for fair rules about unavoidably shared resources. So while we continue to apply market pressures, we can also work to sharpen commonsense regulatory protections for food and drug safety, environmental safeguards, and other best practices for maximizing health and sustainability, for the long term. When choosing what policies to support, our first priority needs to be wholeness – whole swaths of land and all the people who depend on it, whole bodies, whole ecosystems.

What is hiding in our food?

Because obesity is linked to so many health problems, it has been an acknowledged, major public health concern for many years, but the half-hearted prevention efforts so far have focused on only the obvious cause, near the end point: excessive feeding and insufficient exercise. Finding a better answer to obesity and its companion diseases means asking better questions, taking a wider view to include the whole body and as many variables as might play a role, looking for pattern and seeking out the symmetries.

Yes, overeating will usually make you fat, but it now seems clear that there are greater problems with our food supply than just portion sizes. With research indicating that even the same diet as earlier generations leads to a greater likelihood of overweight, today's dialogue about obesity still largely ignores an obvious candidate for attention beyond calorie counting – hormones, the chemical catalysts known to stir up the body, pull on pounds, drive appetites, and change behaviors. And we now know enough about the connections between organisms, even between plants and animals, to consider whether and how not only hormones but growth drivers such as the synthetic nitrates in Miracle-Gro or the antibiotics that mysteriously accelerate chicken growth might be hastening the pace and scale of childhood development, or of adult weight gain, or of cancer growth. With only partial explanations for our epidemics of overgrowth, these are the questions symmetry asks of a new generation

of researchers – to find new patterns of linkage between existing data.

Each molecule of every body part is a building block that entered through the foods we eat, the fluids we drink, or air we breathe, and the greatest of these is food. Not only are we what we eat, but we are what our foods have been eating, whether literally or, for plants, through their soil and water, and the quality of our food is determined by specific and almighty industries. And, of these, the corporations and systems that matter most to our health are those that decide our dietary choices.

We can expect that human ingenuity and molecular engineering will further sharpen the primordial swords to make ever-more potent synthetic hormones, and business will want to use all of the latest tools to drive productivity and increase market share, but we must recognize the interconnectedness of our human lives. Soaring rates of obesity, diabetes, and cancer frame a public health crisis for America and an unmanageable challenge for health care – a catastrophe already begun and looming larger as we look forward. And with obesity no longer a marker of wealth but of poverty, it is the poor who carry the heaviest burden of cheap junk food, as they balloon from obese children to sickened adults.

Still time to correct the mismatch

The human body is a living triumph of evolution, sustained by water and naturally nutrient-dense foods, resistant to ancient challenges of drought and famine, but exquisitely vulnerable to new threats of refined excess, industrial junk foods, and endocrine disruptors. Big Agribusiness, Big Food, Big Soda, and Big Pharma have achieved great financial successes, papering over their dangers with brightly positive images, but each casts its own noxious shadow over human health and wellness.

The timeline of human existence on earth is thought to be around five hundred thousand years, but the new problems of chemical agribusiness, excessive production and use of synthetic nitrates, and artificial hormone abuse have played out in the course of a single human lifetime. Taking the long view, this deviation has lasted for only the briefest moment of human history, and perhaps none of the pernicious consequences will prove irreversible if we act now to change direction.

In the span of one human lifetime, we have also witnessed the rise and fall of antibiotics, at first almost universally effective in treating infections and now limited by excessive use and bacterial resistance. Symmetry reveals the parallel folly of a national agribusiness gamble, where blunt or blind applications of military-industrial surpluses – of synthetic nitrate fertilizers and petrochemical herbicides and insecticides – have driven the postwar commitment to monoculture, at the expense of our soil, water quality and human health and fertility.

Our personal resourcefulness and collective wisdom can correct the course, but those who have gained most – today's corporate giants and titans of industry – are formidable adversaries. Expect little call for change from them, locked in the status quo, nor from the politicians whom they fund, nor from a medical establishment invested in disease. It is up to us. It is our choice of what to buy, what to eat, which production practices to support, and how to vote – where to find our place and make our way in the flock.

Cheap American processed foods, formulated to mop up the excesses of monoculture, fueled by hormones, vigorously marketed, and supported by government subsidies are important factors in the tragedy of today's obesity, heart disease, cancer, and now dementia, as well as the heartbreak of babies born already unable to fulfill the full potential of a healthy human life. It is not enough to look at narrow decisions without the context of the whole. We must insist that all of our collaborative wisdom is brought to bear to broaden the view, on a national and international scale. If we restore the delicate balance of life, we will protect the health of all, from the old to the young to the unborn. As we step back and look at the whole, our challenge is to ask the symmetry-seeking questions.

Index

Note: Page locators followed by f indicate figures.

Y

Z